Micromanipulation

Micromanipulation

Special Issue Editor

Nicola Pio Belfiore

MDPI • Basel • Beijing • Wuhan • Barcelona • Belgrade

MDPI

Special Issue Editor
Nicola Pio Belfiore
Università degli studi di Roma Tre
Italy

Editorial Office
MDPI
St. Alban-Anlage 66
4052 Basel, Switzerland

This is a reprint of articles from the Special Issue published online in the open access journal *Actuators* (ISSN 2076-0825) in 2018 (available at: https://www.mdpi.com/journal/actuators/special_issues/micromanipulation)

For citation purposes, cite each article independently as indicated on the article page online and as indicated below:

LastName, A.A.; LastName, B.B.; LastName, C.C. Article Title. *Journal Name* **Year**, *Article Number*, Page Range.

ISBN 978-3-03897-503-8 (Pbk)
ISBN 978-3-03897-504-5 (PDF)

Cover image courtesy of Alvise Bagolini and Michele Fedrizzi.

Contents

About the Special Issue Editor

Nicola Pio Belfiore, Professor, IEEE Member, teaches "applied mechanics" and "functional design" at the University of Roma Tre, Italy. After obtaining his Ph.D. degree, completed at "Sapienza" in cooperation with the University of Maryland of College Park, he won two international awards in the USA and one in Europe. Since 2008, he has been Honorary Professor of Obuda University, Hungary. The author of three textbooks, two patents, and about one hundred scientific papers, he has been the coordinator of several scientific projects, both National and European. In 2013, he was the director of the Second Level Vocational Master in Energy Conversion Efficiency and Renewable Energy. His interests are functional design, MEMS, robotics, tribology, kinematics, and dynamics. In October 2017, he moved from Sapienza to Roma Tre University.

Preface to "Micromanipulation"

The idea of dedicating a Special Issue to the field of micromanipulation came about from the society's increasing interest in this subject. The demand of microsystems for micromanipulation has grown quite quickly in recent decades, and many fields of applications, nowadays, need new, efficient tools for this peculiar purpose. The problem of actuation is crucial for micromanipulation, and, for this reason, the present book is believed to offer a small contribution to its solution. There is a certain consciousness that this subject must be a highly interdisciplinary one, because the "engineering" offer and the "applications" demand must meet at a certain point. Therefore, the present Issue is composed of papers that are clearly located in the engineering field, although they suggest a variety of implications for many other topics, such as the biomedical field. The purpose of this Special Issue is to stimulate the interest of colleagues, engineers, and students towards a peculiar aspect of engineering with a strong interdisciplinary content: the physical manipulation of micro objects at the micro scale. The development of didactical methods dedicated to this special field has been also covered with two papers. I was very pleased to have received a number of contributions from my colleagues, who agreed to share their results with us, publishing their papers within the present Special Issue. I am very grateful to them. Finally, I would like to thank all the staff of MDPI for supporting and trusting me: my service for "Actuators" has helped me a lot to develop a growing sensitivity towards the world of research and to the dissemination of scientific results. Special thanks is due to Darci Zhou, who has supported me to develop the present Special Issue since February 2018, for her help and enthusiasm.

Nicola Pio Belfiore
Special Issue Editor

actuators

MDPI

Editorial
Micromanipulation: A Challenge for Actuation

Nicola Pio Belfiore

Department of Engineering, Università degli studi di Roma Tre, via della Vasca Navale 79, 00146 Roma, Italy; nicolapio.belfiore@uniroma3.it; Tel.: +39-06-5733-3316

Received: 12 November 2018; Accepted: 29 November 2018; Published: 3 December 2018

Abstract: Manipulating micro objects has become an important task in several applications. Actuation is a crucial aspect of micromanipulation because there are physical restrictions which affect actuators' performances at the micro or nano scale. One way of getting rid of these limitations is the use of an appropriate mechanical structure which enhances the elasticity of the material or provides mechanical advantage. This Special Issue of Actuators, which is dedicated to micromanipulation, offers a contribution to the development of some promising methods to actuate a microsystem for micromanipulation.

Keywords: micromanipulation; nanomanipulation

1. Introduction

During the last decades, microsystems have been developing very fast thanks to progress in science and technology, giving rise to two main crucial questions, namely,

- how could these micro devices be fabricated and operated?
- how could they improve certain aspects of our lives?

The first question leads to classical issues of engineering, such as design, fabrication and control, which study the most convenient way to create the tools for micromanipulation.

The second question is clearly related to the applications of these new developed tools and to their exploitation, as an opportunity to solve old and new problems for the improvement of certain aspects of our lives.

1.1. Downsizing Effects

Once involved in the development of a microsystem, designers immediately come across the problem of handling the scaling effects. They soon become capable of monitoring how surface and volume properties change their impact on a system; once the latter is downsized by one or more orders of magnitude from the human-size typical dimensions: of course, surfaces or volumes will reduce their quantities by the square or the cube, respectively, of the reduced lengths.

Scaling effects were described by Galileo Galilei in 1638 [1]. He pointed out that "a large animal does not possess simply a bone on a larger scale, but its thickness must increase more quickly than the length of the relevant bone", because resistance and weight scale their quantities differently during downsizing. Similarly, "a giant would never have the same limb ratios of a man with a normal size, but he must have thicker limbs suitable to support its mass".

After about a century, Jonathan Swift described the voyage of Lemuel Gulliver [2] to Lilliput. In the book, the imaginary Surgeon, and then a Captain, reports, literally, that the Lilliputian "mathematicians, having taken the height of my body by the help of a quadrant, and finding it to exceed theirs in the proportion of twelve to one, they concluded from the similarity of their bodies, that mine must contain at least 1724 of theirs, and consequently would require as much food as was necessary to support that number of Lilliputian". Surprisingly, the mentioned human-to-lilliputian food ratio (1724) only approximates the square of 12 (1728) and, even more curiously, the year of publication of the book (1726).

1.2. Design, Fabrication, and Control

Design, fabrication and control of microsystems for micromanipulation have to face several difficulties because, at the micro or nano scale levels, several paradigms of macro-scale engineering are no longer valid [3]. These and many other consequences and implications have been early underlined by Richard P. Feynman, during his seminal talk given on 26 December 1959 at the annual meeting of the American Physical Society (APS) at the California Institute of Technology [4]. In this speech, Feynman surprisingly mentioned many applications such as computer miniaturization, microsurgery, micro-machining and actuation.

Nowadays, an increasing number of applications are demanding high-precision tools which introduce more and more constraints to their design, following the classical client-to-designer feedback.

The following issues have been particularly discussed in the literature.

- Micromanipulation mechanics [5–7]:

  theoretical modeling,
  numerical simulation, and
  experimental testing.

- Microsystems architecture, components and manufacturing [8–12]:

  design,
  fabrication,
  fabrication constrained design rules,
  packaging and
  biocompatibility.

- Actuation and Sensing [13–16]:

  electrostatic,
  electrothermal,
  electromagnetic,
  piezoelectric.

- Micro-electro-mechanical system (MEMS) integration:

  lab-on-chip systems [17–19],
  MEMS integration with Application Specific Integrated Circuit (ASIC) [20,21].

- Control [22–24]:

  automatic regulation and control of microsystems,
  operational aspects of micromanipulation,
  measurements.

However, while Nanotechnology has found a certain variety of good (although still perfectible) solutions to many problems in MEMS developing [25], what makes micromanipulation particularly difficult is its actuation. In fact, the available sources are often not able to exert an adequate force or torque. Furthermore, the highest performances are achieved only within a rather restricted range for end-effector displacements, and so the mechanical structure of the microsystem must be optimized. In fact, actuators are devices which transform energy (e.g., thermal or electrical, depending on the available source) in mechanical energy. Using the classical sources, such as electrostatic, thermal (shape memory, electrothermal), electromagnetic, or piezoelectric [26], it turns out that it is rather difficult to achieve both a large force together with a large displacement at the output link and, therefore, a micro mechanism can be employed to gain the mechanical advantage. Fortunately, recent developments in MEMS Technology allows designers to introduce different sorts of micro mechanisms, such as microgrippers [27,28], with multi-hinge and multi-DoF (Degrees of Freedom) properties, and other multi-axes devices [29]. Such opportunity gave rise to the introduction of a design technique based on the rigid-body replacement method [30], which refers to classical issues of mechanism science such as topology [31,32], kinematic synthesis [33–35], kinetostatic indexing [36], isotropic compliance [37,38] and parametric design [39,40].

1.3. Applications

Nowadays, micro or nano manipulation is very attractive in a large variety of applications, ranging from medicine, surgery or biology, to microelectronics, micro mechanics and aerospace. Therefore, the following items represent a non-exhaustive list of much more examples of applications:

- drug delivery [41,42],
- minimal-invasive surgery [43,44],
- tissue or cell manipulation [30,45–47],
- diagnostics [48–51],
- aerospace [52–54],
- micro-assembly [16,55–57], and
- microelectronics [58,59].

1.4. Forthcoming or Emerging Issues

Other than the above-mentioned classical issues, there are some others that are either more recent or, at least, less usual than the former. In fact, once new concepts become well-established and real devices, the interest in their optimization and use increases. Furthermore, some other related issues appear, and so the following topics could soon become new topics for MEMS and microsystem applications.

- Computational intelligence: optimization and control of microsystems [60,61].
- Development of ambient intelligence [62] based on sensors, actuators and standardized internet communication.
- Use of nonlinearity benefits [63,64].
- Configuration management and reconfigurable manufacturing systems for the development of microsystems during lifecycle [65].
- Ethics: ethical issues in the activities of criteria-based decision making autonomous micro-manipulators in the medical, biological, aerospace and industrial fields [66–70].
- Education: new trends in microsystems teaching–learning methods, tinkering, open access, wiki tools [71,72].

2. On the Variety of Demand vs. the Supplying Capability

A selection of 1846 papers concerning "micromanipulation", distributed over seven different meta categories, namely, Biology, Computer Science, Engineering, Medicine, Multidisciplinary issues, Physics/Chemistry, Science have been analyzed. These categories have been named after the classification suggested by one of the most acknowledged database for Science and Technology, that is the Web of Science [73]. A series of queries has been launched on the database and some statistical data have been extracted. At the first sight, among the above mentioned categories, there is one which can be assumed to represent the "supplying capacity" at the actual state of the art, as introduced in Section 1.2, namely,

- Engineering,

The other six categories represent the variety of the demand of micromanipulation technology and its temporary success in applications, as mentioned in Section 1.3, that are

- Biology,
- Physics and Chemistry,
- Medicine,
- Computer Science,
- Multidisciplinary issues, and
- Science.

Statistical analysis has revealed that the struggling for technological readiness, that could be reasonably represented by the filed of engineering, collects almost a half of all contributions, about 45% of the papers. On the other hand, the applications, which could be thought as the "recipient" of the technological progress, gather the other half of the full bunch of papers. More in detail, the Biomedical Sciences, including Biology (18%) and Medicine (11%), form 29% of contributions. General categories such as Science (3%) and Physics/Chemistry (13%) form about 16% of the considered papers. Finally, Computer Science includes 6% of the selected papers, while the last 4% concerns multidisciplinary issues. Figure 1 illustrates the distribution of the selected papers according to the above-mentioned categories.

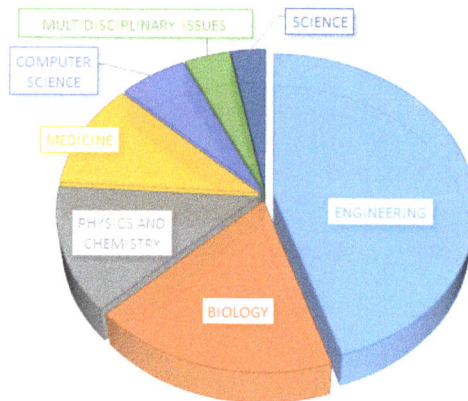

Figure 1. Statistical distribution of 1864 papers concerning "micromanipulation" along seven different groups, gathered over 83 Web of Science Categories.

The adopted method of paper selection has been based on classical database query tools, as implemented in WoS [73] and, therefore, the search keys, which consist of selected words, may be subject to double meaning errors. This introduces an error that can be roughly estimated by manually checking some elements extracted from a randomly sampled group. On the basis of a rough estimation, the mistake is expected not to exceed 3% of the values.

The results of the present investigation show that research into micromanipulation is still more extended in the field of engineering than in any of the other applicative fields. However, the full group of applications represents half of the analyzed papers, which shows that engineering vs. applications are nowadays in balance.

These results probably reveal the actual struggle of investigators to improve the technical characteristics from a rather low technological readiness level (TRL) [74], to higher TRL values that are more suitable for immediate and commercial applications. It is rather difficult to predict how much time it takes to make micromanipulation technology ready for more applications in hospitals, labs, cars, aircraft, appliances, and so on, while, on the other hand, it is easy to infer that the number of applications will increase as much as new confidence is acquired in the construction and control of the tools.

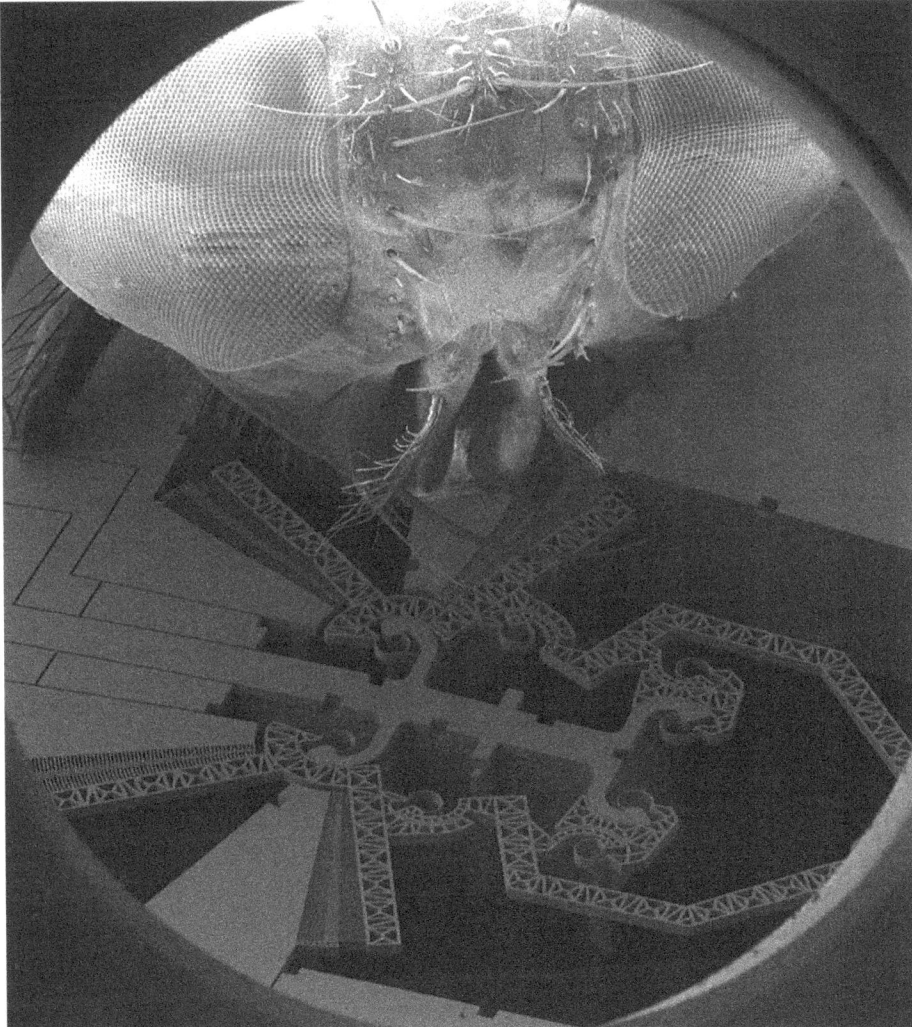

Figure 2. SEM image obtained by Alvise Bagolini and Michele Fedrizzi at Micro Nano Fabrication and Characterization Facility, Fondazione Bruno Kessler, Trento, Italy (PS the housefly has been found dead in Villamontagna, Trento, Italy and no harm has been inflicted to the poor insect).

3. Tools for the Observation

As known, the human eye normally barely distinguish objects within the size of a few tens of microns, which makes it difficult for us to monitor micro devices without a proper means of magnification. Usually, optical microscopy could give a first-hand tool for observation, with the advantage of operating in air or liquid environments. However, SEM observation is more helpful to characterize the prototypes, because of their higher resolution. For this reason, in MEMS development, the moment always comes, eventually, to take an SEM picture of a microsystem. Usually, an image obtained by an SEM instrument declares the adopted magnification factor and other technical data and, therefore, there is no real need to use a side comparison object (whose dimensions are universally known). At the macro level, classical objects can be used to compare the observed subject with familiar objects, such as a one cent coin or the portion of a ruler. However, while moving around either the "brobdingnag" or the "lilliput" worlds, it is not always easy to grasp the real sense of the size of an object which has unusual dimensions, such as gigantic (planets, stars, galaxies) or minuscule (cells, atoms) things.

Figure 2 is an example of a comparison exercise at the microscale. An adult housefly, musca domestica, was positioned over the square portion of a silicon wafer where a microgripper had been fabricated [12]. According to the construction process and the adopted parameters, each microgripper lays within a 5×5 mm^2 window, whereas adult houseflies are usually from 6 to 7 mm in size. That is why the whole insect was as long as the length of a microgripper window. The fly eyes, whose diameter is approximately one millimeter and a half, are visible at the top of Figure 2, while the microgripper shows up in the lower half of the picture. In the upper half of the figure, the compound eyes of the fly appear in all their complexity. In fact, there are almost 3500 ommatidia facets in the musca domestica' eye [75], while the single ommatidium is about 10 μm [76]. This figure make the microgripper a bit less mysterious object to our understanding.

4. Micro- or Nano-Manipulators vs. Micro- or Nano-Robots

Finally, it is worth noting that the terms micromanipulator and microrobot (the same holds for the pair nanomanipulator and nanorobot) are often used as synonymous, while it would be better not to confuse them that way. In fact, developments in technology have made it possible to build devices with different features; for example, their overall size or their positioning or manipulating accuracy.

Considering the scope of developing miniaturized devices at the bottom scale, it seems to be appropriate to refer to such devices as microrobots, micromechanisms, micromachines or, more generally, microsystems (nanorobots, nanomechanism, and so on). On the other hand, when a device (no matter about its size) is required to be able to manipulate objects with micro (or nano) accuracy, then it seems that the term micromanipulators (or nanomanipulators) becomes more appropriate.

This distinction is quite substantial. In fact, in many occasions, it is necessary to handle micro- or nano-objects, with limited concern about the size of the manipulating object. For example, nano-manipulators used inside the chamber of a Scanning Electron Microscope (SEM) may have extraordinary resolution, <0.5 nm, with rather a, relatively, large overall size, of about 6 cm [77]. On the other hand, in some other circumstances, it is necessary to have a small system, namely, a microsystem, which is able to operate within specific environments, with very restrictive size limits. Interest in miniaturizing micromechanisms [78], microrobots [79,80], microgrippers [81,82], and microtribometer [83] has been recently expressed. In fact, in all these examples. the size of the microsystem was fundamental for the performance of its function.

5. Conclusions

The present Editorial has been written with the intent of intriguing readers and colleagues who may want to deepen the topic of actuation at the microscale, specially for the purpose of micromanipulation. The actual state-of-the-art shows that this problem is rather far from being exhausted and, therefore, the present Special Issue represents an invitation to accept the challenge to find proper solutions to this endeavour.

Conflicts of Interest: The author declears no conflict of interest.

References

1. Galilei, G. *Discourses and Mathematical Demonstrations Relating to Two New Sciences [In Italian: Discorsi e Dimostrazioni Matematiche Intorno a Due Nuove Scienze Attenenti Alla Meccanica E I Movimenti Locali]*; Lodewijk Elzevir: Leiden, The Netherlands, 1638.
2. Swift, J. *Travels into Several Remote Nations of the World. In Four Parts. By Lemuel Gulliver, First a Surgeon, and then a Captain of Several Ships*; Benjamin Motte: London, UK, 1726.
3. Bhushan, B.E. (Ed.) *Springer Handbook of Nanotechnology*; Springer: Berlin/Heidelberg, Germany, 2017.
4. Feynman, R.P. There's plenty of room at the bottom [data storage]. *J. Microelectromech. Syst.* **1992**, *1*, 60–66. [CrossRef]
5. Huang, H.; Kamm, R.D.; Lee, R.T. Cell mechanics and mechanotransduction: Pathways, probes, and physiology. *Am. J. Physiol. Cell Physiol.* **2004**, *287*, C1–C11. [CrossRef] [PubMed]
6. Swaminathan, V.; Mythreye, K.; Tim O'Brien, E.; Berchuck, A.; Blobe, G.; Superfine, R. Mechanical Stiffness grades metastatic potential in patient tumor cells and in cancer cell lines. *Cancer Res.* **2011**, *71*, 5075–5080. [CrossRef] [PubMed]
7. Zhang, H.; Liu, K.K. Optical tweezers for single cells. *J. R. Soc. Interface* **2008**, *5*, 671–690. [CrossRef]
8. Potrich, C.; Lunelli, L.; Bagolini, A.; Bellutti, P.; Pederzolli, C.; Verotti, M.; Belfiore, N.P. Innovative silicon microgrippers for biomedical applications: Design, mechanical simulation and evaluation of protein fouling. *Actuators* **2018**, *7*, 12. [CrossRef]
9. Huesgen, T.; Woias, P.; Kockmann, N. Design and fabrication of MEMS thermoelectric generators with high temperature efficiency. *Sens. Actuators A Phys.* **2008**, *145–146*, 423–429. [CrossRef]
10. Judy, J.W. Microelectromechanical systems (MEMS): Fabrication, design and applications. *Smart Mater. Struct.* **2001**, *10*, 1115–1134. [CrossRef]
11. Shen, D.; Park, J.H.; Ajitsaria, J.; Choe, S.Y.; Wikle, H.C.; Kim, D.J. The design, fabrication and evaluation of a MEMS PZT cantilever with an integrated Si proof mass for vibration energy harvesting. *J. Micromech. Microeng.* **2008**, *18*, 055017. [CrossRef]
12. Bagolini, A.; Ronchin, S.; Bellutti, P.; Chistè, M.; Verotti, M.; Belfiore, N.P. Fabrication of Novel MEMS Microgrippers by Deep Reactive Ion Etching With Metal Hard Mask. *J. Microelectromech. Syst.* **2017**, *26*, 926–934. [CrossRef]
13. Rezazadeh, G.; Tahmasebi, A.; Zubstov, M. Application of piezoelectric layers in electrostatic MEM actuators: Controlling of pull-in voltage. *Microsyst. Technol.* **2006**, *12*, 1163–1170. [CrossRef]
14. Pacheco, S.P.; Katehi, L.P.B.; Nguyen, C.T.C. Design of low actuation voltage RF MEMS switch. In Proceedings of 2000 IEEE MTT-S International Microwave Symposium Digest, Boston, MA, USA, 11–16 June 2000; pp. 165–168.
15. Bell, D.J.; Lu, T.J.; Fleck, N.A.; Spearing, S.M. MEMS actuators and sensors: Observations on their performance and selection for purpose. *J. Micromech. Microeng.* **2005**, *15*, S153–S164. [CrossRef]
16. Donald, B.R.; Levey, C.G.; Paprotny, I. Planar microassembly by parallel actuation of MEMS microrobots. *J. Microelectromech. Syst.* **2008**, *17*, 789–808. [CrossRef]
17. Andrieux, G.; Eloy, J.C.; Mounier, E. Technologies and market trends for polymer MEMS in microfluidics and lab-on-chip. In Proceedings the of SPIE, Progress in Biomedical Optics and Imaging, San Jose, CA, USA, 22–27 January 2005; pp. 60–64.
18. De Pasquale, G.; Bertana, V.; Scaltrito, L. Experimental evaluation of mechanical properties repeatability of SLA polymers for labs-on-chip and bio-MEMS. *Microsyst. Technol.* **2018**, *24*, 3487–3497. [CrossRef]
19. Palmieri, M. DNA lab on chip rests on MEMS foundation. *EDN* **2004**, *49*, 24–28.
20. Takahashi, K.; Kwon, H.N.; Mita, M.; Fujita, H.; Toshiyoshi, H.; Suzuki, K.; Funaki, H. Monolithic integration of high voltage driver circuits and MEMS actuators by ASIC-like postprocess. In Proceedings of the International Conference on Solid State Sensors and Actuators and Microsystems, Digest of Technical Papers, TRANSDUCERS '05, Seoul, Korea, 5–9 June 2005; Volume 1, pp. 417–420.
21. Grayver, E.; M'Closkey, R.T. Automatic gain control ASIC for MEMS gyro applications. In Proceedings of the American Control Conference, Arlington, VA, USA, 25–27 June 2001; Volume 2, pp. 1219–1222.
22. Borovic, B.; Liu, A.Q.; Popa, D.; Cai, H.; Lewis, F.L. Open-loop versus closed-loop control of MEMS devices: Choices and issues. *J. Micromech. Microeng.* **2005**, *15*, 1917–1924. [CrossRef]

23. Park, S.; Horowitz, R. Adaptive control for the conventional mode of operation of MEMS gyroscopes. *J. Microelectromech. Syst.* **2003**, *12*, 101–108. [CrossRef]

24. Zheng, Q.; Dong, L.; Lee, D.H.; Gao, Z. Active disturbance rejection control for MEMS gyroscopes. *IEEE Trans. Control Syst. Technol.* **2009**, *17*, 1432–1438. [CrossRef]

25. Gad-el Hak, M.E. (Ed.) *The MEMS Handbook*, 2nd ed.; Mechanical and Aerospace Engineering Series; CRC Press, Tayor and Francis Group: Boca Raton, FL, USA, 2005.

26. Dochshanov, A.; Verotti, M.; Belfiore, N.P. A Comprehensive Survey on Microgrippers Design: Operational Strategy. *J. Mech. Des. Trans. ASME* **2017**, *139*, 070801. [CrossRef]

27. Belfiore, N.P.; Broggiato, G.; Verotti, M.; Balucani, M.; Crescenzi, R.; Bagolini, A.; Bellutti, P.; Boscardin, M. Simulation and construction of a MEMS CSFH based microgripper. *Int. J. Mech. Control* **2015**, *16*, 21–30.

28. Belfiore, N.P.; Verotti, M.; Crescenzi, R.; Balucani, M. Design, optimization and construction of MEMS-based micro grippers for cell manipulation. In Proceedings of the ICSSE 2013 IEEE International Conference on System Science and Engineering, Budapest, Hungary, 4–6 July 2013; pp. 105–110.

29. Marano, D.; Cammarata, A.; Fichera, G.; Sinatra, R.; Prati, D. Modeling of a Three-Axes MEMS Gyroscope with Feedforward PI Quadrature Compensation; In *Advances on Mechanics, Design Engineering and Manufacturing*; Eynardetal, B., Ed.; Springer: Cham, Switzerland, 2017; pp. 71–80.

30. Sanò, P.; Verotti, M.; Bosetti, P.; Belfiore, N.P. Kinematic Synthesis of a D-Drive MEMS Device with Rigid-Body Replacement Method. *J. Mech. Des. Trans. ASME* **2018**, *140*, 075001. [CrossRef]

31. Belfiore, N.P. Distributed Databases for the development of Mechanisms Topology. *Mech. Mach. Theory* **2000**, *35*, 1727–1744. [CrossRef]

32. Belfiore, N.P. Brief note on the concept of planarity for kinematic chains. *Mech. Mach. Theory* **2000**, *35*, 1745–1750. [CrossRef]

33. Pennestrì, E.; Belfiore, N.P. On the numerical computation of Generalized Burmester Points. *Meccanica* **1995**, *30*, 147–153. [CrossRef]

34. Pennestrì, E.; Belfiore, N.P. Modular third-order analysis of planar linkages with applications. In *ASME Design Technical Conference, Mechanism Synthesis and Analysis*; ASME: New York, NY, USA, 1994; Volume 70, pp. 99–103.

35. Pennestrì, E.; Belfiore, N.P. On Crossley's contribution to the development of graph based algorithms for the analysis of mechanisms and gear trains. *Mech. Mach. Theory* **2015**, *89*, 92–106. [CrossRef]

36. Sinatra, R.; Cammarata, A.; Angeles, J. Kinetostatic and inertial conditioning of the McGill schnflies-motion generator. *Adv. Mech. Eng.* **2010**, *2*, 186203.

37. Verotti, M.; Belfiore, N.P. Isotropic compliance in E(3): Feasibility and workspace mapping. *J. Mech. Rob.* **2016**, *8*, 061005. [CrossRef]

38. Verotti, M.; Masarati, P.; Morandini, M.; Belfiore, N.P. Isotropic compliance in the Special Euclidean Group SE(3). *Mech. Mach. Theory* **2016**, *98*, 263–281. [CrossRef]

39. Kwak, B.M.; Haug, E.J. Optimal synthesis of planar mechanisms by parametric design techniques. *Eng. Optim.* **1976**, *2*, 55–63. [CrossRef]

40. Lu, Q.; Huang, W.; Sun, M. Parametric design of flexible amplification mechanism based on flexure hinge. *J. Vib. Meas. Diagn.* **2016**, *36*, 935–941.

41. Paul, S.R.; Nayak, S.K.; Anis, A.; Pal, K. MEMS-Based Controlled Drug Delivery Systems: A Short Review. *Polym. Plast. Technol. Eng.* **2016**, *55*, 965–975. [CrossRef]

42. Lee, H.J.; Choi, N.; Yoon, E.S.; Cho, I.J. MEMS devices for drug delivery. *Adv. Drug Deliv. Rev.* **2018**, *128*, 132–147. [CrossRef] [PubMed]

43. Rebello, K.J. Applications of MEMS in surgery. *Proc. IEEE* **2004**, *92*, 43–55. [CrossRef]

44. Park, Y.S.; Gopalsami, N.; Gundeti, M.S. Tactile MEMS-based sensor element for robotic surgery. In Proceedings of the American Nuclear Society 2014 Annual Meeting, Reno, NV, USA, 15–19 June 2014; pp. 43–44.

45. Chronis, N.; Lee, L.P. Polymer mems-based microgripper for single cell manipulation. In Proceedings of the Seventeenth IEEE International Conference on Micro Electro Mechanical Systems, Maastricht, The Netherlands, 25–29 January 2004; pp. 17–20.

46. Pan, P.; Wang, W.; Ru, C.; Sun, Y.; Liu, X. MEMS-based platforms for mechanical manipulation and characterization of cells. *J. Micromech. Microeng.* **2017**, *27*, 123003. [CrossRef]

47. Di Giamberardino, P.; Bagolini, A.; Bellutti, P.; Rudas, I.J.; Verotti, M.; Botta, F.; Belfiore, N.P. New MEMS tweezers for the viscoelastic characterization of soft materials at the microscale. *Micromachines* **2017**, *9*, 15. [CrossRef] [PubMed]
48. Polla, D.L.; Krulevitch, P.; Wang, A.; Smith, G.; Diaz, J.; Mantell, S.; Zhou, J.; Zurn, S.; Nam, Y.; Cao, L.; Hamilton, J.; Fuller, C.; Gascoyne, P. MEMS-based diagnostic microsystems. In Proceedings of the 1st Annual International IEEE-EMBS Special Topic Conference on Microtechnologies in Medicine and Biology, Lyon, France, 12–14 October 2000; pp. 41–44.
49. Huang, Y.; Mather, E.L.; Bell, J.L.; Madou, M. MEMS-based sample preparation for molecular diagnostics. *Fresenius' J. Anal. Chem.* **2002**, *372*, 49–65. [CrossRef] [PubMed]
50. Gnerlich, M.; Perry, S.F.; Tatic-Lucic, S. A submersible piezoresistive MEMS lateral force sensor for a diagnostic biomechanics platform. *Sens. Actuators A Phys.* **2012**, *188*, 111–119. [CrossRef]
51. Pandya, H.J.; Park, K.; Chen, W.; Goodell, L.A.; Foran, D.J.; Desai, J.P. Toward a Portable Cancer Diagnostic Tool Using a Disposable MEMS-Based Biochip. *IEEE Trans. Biomed. Eng.* **2016**, *63*, 1347–1353. [CrossRef]
52. Ho, C.M.; Tung, S.; Lee, G.B.; Tai, Y.C.; Jiang, F.; Tsao, T. MEMS—A technology for advancements in aerospace engineering. In Proceedings of the 35th Aerospace Sciences Meeting and Exhibit, Reno, NV, USA, 6–9 January 1997.
53. Kraft, M.; White, N.M. *MEMS for Automotive and Aerospace Applications*; Woodhead Publishing Series in Electronic and Optical Materials, Elsevier Ltd.: Amsterdam, The Netherlands, 2013; pp. 1–342.
54. Bhat, K.N.; Nayak, M.M.; Kumar, V.; Thomas, L.; Manish, S.; Thyagarajan, V.; Gaurav, S.; Bhat, N.; Pratap, R. Design, development, fabrication, packaging, and testing of MEMS pressure sensors for aerospace applications. In *Micro and Smart Devices and Systems*; Vinoy, K.J., Ananthasuresh, G.K., Pratap, R., Krupanidhi, S.B., Eds.; Springer: New Delhi, India, 2014; Volume 14, pp. 3–17.
55. Nelson, B.J.; Zhou, Y.; Vikramaditya, B. Sensor-Based Microassembly of Hybrid MEMS Devices. *IEEE Control Syst.* **1998**, *18*, 35–45.
56. Tsui, K.; Geisberger, A.; Ellis, M.; Skidmore, G. Micromachined end-effector and techniques for directed MEMS assembly. *J. Micromech. Microeng.* **2004**, *14*, 542–549. [CrossRef]
57. Wei, Y.; Xu, Q. An overview of micro-force sensing techniques. *Sens. Actuators A Phys.* **2015**, *234*, 359–374. [CrossRef]
58. Lin, L. MEMS post-packaging by localized heating and bonding. *IEEE Trans. Adv. Packag.* **2000**, *23*, 608–616. [CrossRef]
59. Howlader, M.M.R.; Okada, H.; Kim, T.H.; Itoh, T.; Suga, T. Wafer level surface activated bonding tool for MEMS packaging. *J. Electrochem. Soc.* **2004**, *151*, G461–G467. [CrossRef]
60. May, G. Intelligent SOP manufacturing. *IEEE Trans. Adv. Packag.* **2004**, *27*, 426–437. [CrossRef]
61. Belfiore, N.P.; Rudas, I. Applications of computational intelligence to mechanical engineering. In Proceedings of the 15th IEEE International Symposium on Computational Intelligence and Informatics, Budapest, Hungary, 19–21 November 2014; pp. 351–368.
62. Delsing, J.; Lindgren, P. Sensor communication technology towards ambient intelligence. *Meas. Sci. Technol.* **2005**, *16*, R37. [CrossRef]
63. Gammaitoni, L.; Neri, I.; Vocca, H. The benefits of noise and nonlinearity: Extracting energy from random vibrations. *Chem. Phys.* **2010**, *375*, 435–438. [CrossRef]
64. Green, P.; Worden, K.; Atallah, K.; Sims, N. The benefits of Duffing-type nonlinearities and electrical optimisation of a mono-stable energy harvester under white Gaussian excitations. *J. Sound Vib.* **2012**, *331*, 4504–4517. [CrossRef]
65. Puik, E.; Gielen, P.; Telgen, D.; van Moergestel, L.; Ceglarek, D. A generic systems engineering method for concurrent development of products and manufacturing equipment. *IFIP Adv. Inf. Commun. Technol.* **2014**, *435*, 139–146.
66. Ailinger, R.L.; Black, P.L.; Lima-Garcia, N. Use of electronic monitoring in clinical nursing research. *Clin. Nurs. Res.* **2008**, *17*, 89–97. [CrossRef]
67. Morgan, D. Respect for autonomy: Is it always paramount? *Nurs. Ethics* **1996**, *3*, 118–125. [CrossRef]
68. Tuma, J.R. Nanoethics in a Nanolab: Ethics via Participation. *Sci. Eng. Ethics* **2013**, *19*, 983–1005. [CrossRef]
69. Makarczuk, T.; Matin, T.R.; Karman, S.B.; Diah, S.Z.M.; Davaji, B.; MacQueen, M.O.; Mueller, J.; Schmid, U.; Gebeshuber, I.C. Biomimetic MEMS to assist, enhance and expand human sensory perceptions—A survey on state-of-the art developments. In *Smart Sensors, Actuators, and MEMS V*; Schmid, A., Sánchez-Rojas, J.L., Leester-Schaedel, M., Eds.; SPIE Digital Library: Bellingham, WA, USA, 2011; Volume 8066.

70. Simou, P.; Alexiou, A.; Tiligadis, K. Artificial humanoid for the elderly people. *Adv. Exp. Med. Biol.* **2015**, *821*, 19–27. [PubMed]

71. Bonciani, G.; Biancucci, G.; Fioravanti, S.; Valiyev, V.; Binni, A. Learning micromanipulation, Part 2: Term projects in practice. *Actuators* **2018**, *7*, 56. [CrossRef]

72. Biancucci, G.; Bonciani, G.; Fioravanti, S.; Binni, A.; Lucchese, F.; Matrisciano, A. Learning micromanipulation, Part 1: An approach based on multidimensional ability inventories and text mining. *Actuators* **2018**, *7*, 55. [CrossRef]

73. Web of Science by Clarivate Analytics, 2018. Available online: https://clarivate.com/products/web-of-science/ (accessed on 18 September 2018).

74. Technology Readiness Levels (TRL), Horizon 2020—Work Programme 2018–2020 General Annexes, Extract from Part 19—Commission Decision C(2017)7124, 2017. Available online: http://ec.europa.eu/research/participants/data/ref/h2020/other/wp/2018-2020/annexes/h2020-wp1820-annex-g-trl%5C_en.pdf (accessed on 16 September 2018).

75. Sukontason, K.L.; Chaiwong, T.; Piangjai, S.; Upakut, S.; Moophayak, K.; Sukontason, K. Ommatidia of blow fly, house fly, and flesh fly: Implication of their vision efficiency. *Parasitol. Res.* **2008**, *103*, 123–131. [CrossRef] [PubMed]

76. Barlow, H.B. The Size of Ommatidia in Apposition Eyes. *J. Exp. Biol.* **1952**, *29*, 667–674.

77. Kleindiek Nanotechnik GmbH. MM3A-EM Micromanipulator, Version 10.01. © Kleindiek Nanotechnik GmbH, 2018. Available online: https://www.nanotechnik.com/fileadmin/public/brochures/mm3a-em.pdf (accessed on 4 September 2018).

78. Belfiore, N.P.; Simeone, P. Inverse kinetostatic analysis of compliant four-bar linkages. *Mech. Mach. Theory* **2013**, *69*, 350–372. [CrossRef]

79. Belfiore, N.P.; Emamimeibodi, M.; Verotti, M.; Crescenzi, R.; Balucani, M.; Nenzi, P. Kinetostatic optimization of a MEMS-based compliant 3 DOF plane parallel platform. In Proceedings of the ICCC 2013 IEEE 9th International Conference on Computational Cybernetics, Tihany, Hungary, 8–10 July 2013; pp. 261–266.

80. Balucani, M.; Belfiore, N.P.; Crescenzi, R.; Verotti, M. The development of a MEMS/NEMS-based 3 D.O.F. compliant micro robot. *Int. J. Mech. Control* **2011**, *12*, 3–10.

81. Verotti, M.; Dochshanov, A.; Belfiore, N.P. Compliance Synthesis of CSFH MEMS-Based Microgrippers. *J. Mech. Des. Trans. ASME* **2017**, *139*, 022301. [CrossRef]

82. Verotti, M.; Dochshanov, A.; Belfiore, N.P. A Comprehensive Survey on Microgrippers Design: Mechanical Structure. *J. Mech. Des. Trans. ASME* **2017**, *139*, 060801. [CrossRef]

83. Belfiore, N.P.; Prosperi, G.; Crescenzi, R. A simple application of conjugate profile theory to the development of a silicon micro tribometer. In Proceedings of the ASME 2014 12th Biennial Conference on Engineering Systems Design and Analysis, Copenhagen, Denmark, 25–27 June, 2014; Volume 2.

actuators

MDPI

Article

Mechanical Response of Four-Bar Linkage Microgrippers with Bidirectional Electrostatic Actuation

Fabio Botta [1,*,†], Matteo Verotti [2], Alvise Bagolini [3], Pierluigi Bellutti [3] and Nicola Pio Belfiore [1]

[1] Department of Engineering, Università degli studi di Roma Tre, 00146 Roma, Italy; fabio.botta@uniroma3.it
[2] Università degli studi Niccolò Cusano, 00166 Roma, Italy; matteo.verotti@unicusano.it
[3] Micro Nano Fabrication and Characterization Facility, Fondazione Bruno Kessler, 38123 Trento, Italy; bagolini@fbk.eu (A.B.); bellutti@fbk.eu (P.B.)
*   Correspondence: fabio.botta@uniroma3.it; Tel.: +36-06-5733-3491
†   Current address: via della Vasca Navale 79, 00146 Rome, Italy.

Received: 2 October 2018; Accepted: 7 November 2018; Published: 11 November 2018

Abstract: This paper presents both an experimental and a numerical study concerning the mechanical response of a silicon microgripper with bidirectional electrostatic actuation to externally applied excitations. The experimental set-up is composed of a probe station equipped with mobile probes that apply contact forces. This part of the investigation aims to test the device's mechanical resistance, its mobility capability and possible internal contacts during the system deformation. The second part of the paper is dedicated to the study of the free undamped vibrations of the microsystem. Finite Element Analysis (FEA) is carried out to evaluate the system vibration modes. The analysis of the modes are useful to predict possible mechanical interference among floating and anchored fingers of the actuating comb drives.

Keywords: MEMS; vibration modes; DRIE; microgripper; comb drive actuators

---

1. Introduction

The introduction of new classes of flexure hinges [1] and the technological progress in mechanical components of MEMS (micro electro mechanical system) [2,3] gave rise to new devices for the manipulation at the microscale. In fact, about a hundred microgrippers [4,5] were designed and fabricated with different purposes and actuation systems. For example, micromanipulation finds important applications in micro assembly processes. Some devices can be fabricated as monolithical structures, whereas others require an assembly step because of particular geometries or different materials [6]. Microgrippers are also employed in optical fibers assembly [7,8]. Another important field of application is biology: manipulation of single cells is an essential step to understand cells behaviors and interactions [9,10]. For example, microgrippers with force sensors where developed for manipulating biological cells [11,12] or to characterize the mechanical properties of biosamples [13,14]. The sensing [15,16] and the control [17,18] of the gripping forces has also been a fundamental issue in developing microgrippers.

However, in spite of the recent progress in nano and micro-machining, there is still a certain difficulty in building multi-hinge and multi-DoF (Degrees of Freedom) MEMS.

The main problem consists of the fact that mobility is granted by flexure hinges and that the latter are still rather complicated to be designed and fabricated in a small portion of the device. This problem is particularly arduous in the design of microgripper for micro manipulation and, therefore, new microsystems equipped with Conjugate Surface Flexure Hinges (CSFHs) have been developed [19] and fabricated [20]. The idea, which dates back to 2012 [21], is based on the partitioning of a block into either rigid or flexible sub-parts, with mobility being granted by the presence of the flexible parts (flexure). A CSFH is a particular kind of flexure that is made of a curved beam, which provide compliance, and a portion of a conjugate-profiles, which provides accuracy. The two components are designed in such a way that the center of the elastic weights of the curved beam is coincident with the center of the conjugate profiles.

The CSFH had a certain number of applications in microsystems, for example in micro mechanisms [22,23], micromanipulators [24–26], tribometers [27], grippers [13], biomechanics, etc. Some recent experimental investigations showed also that these microsystems, although with some restrictions, can be operated by means of comb drives [28]. However, there are still some concerns regarding the feasibility and robustness of these devices in both static and dynamic conditions.

Actuation is among the most important functions of a microsystem, and so different kinds of solutions to this problem have been proposed, such as electrothermal [29,30], shape memory alloy [31,32], or piezoelectric actuators [33,34]. Electrostatic actuators, in particular linear [35–39] or rotary [40–42] comb drives, offer also a feasible actuation system for micromanipulation.

In this investigation, to evaluate the mechanical robustness of the microgripper towards its application in an operational environment, the mechanical functionality of a four-bar linkage microgripper with bidirectional electrostatic actuation has been examined both in static and dynamic conditions. Static load was experimentally tested within a probe station. Then, the static response of the microgripper under externally applied forces has been numerically simulated by means of Finite Element Analysis (FEA) and some original design charts have been built to predict possible contacts between CSFH conjugate surfaces. This information is useful in the design steps, to optimize the orientation of the four CSFHs embedded in the four-bar linkage. Finite Element Analysis (FEA) has been also applied to analyze the vibration modes. The eigenmodes and eigenfrequencies have been calculated and the shapes of vibration associated with the first six modes have been analyzed.

2. Design

The MEMS consists of a bulk structure which gains mobility thanks to elasticity [43]. Generally, compliant mechanisms can be categorized into two main classes: with lumped or distributed compliance. The CSFHs are particularly suitable to be used as elements of compliant mechanisms with lumped compliance. This opportunity gives rise to new design methods which make use of a topological approach [44,45], such as for example, the rigid-body replacement method [46,47]. By following this approach, starting from a classic four-bar linkage, the mask represented in Figure 1a has been created in a configuration that maximizes the comb drives' rotations (see the comb drive detailed view presented in Figure 1b) in both closure and opening directions. Then, one CSFH replaces each rotational joint, as illustrated in the detailed view of Figure 1c. Figure 1d shows the gap Figure 1e between the conjugate surfaces in a non contact configuration.

Once each one of the four ordinary revolute pairs has been replaced by a CSFH, a fully compliant four-bar linkage is created (lumped elasticity). Of course, two compliant four-bar mechanisms are needed to assure the grasping operation and, therefore, two four-bar mechanisms are symmetrically positioned to allow the gripping jaws to symmetrically approach the micro object. With reference to Figure 1, the design illustrated by means of the mask Figure 1a has been laid out in such a way that a gripping jaw is attached to the coupler link. For example, the left jaw is pointed out in Figure 1f. The four-bar mechanism provides the jaw tip a wide range of motion from the open to the close extreme configurations.

The device is operated by means of two bidirectional electrostatic actuators. The open position of the jaws is obtained by applying a voltage between the pads (i) and (ii) represented in Figure 2a. With reference to the Figure 2b, the mobile set of fingers (iv) rotates counterclockwise, while the coupler link rotates clockwise because of the given configuration of the four-bar linkage. By applying a voltage between pads (i) and (iii) of Figure 2a, the opposite applies, and the comb drive mobile wing (v) rotates clockwise, inducing a counterclockwise rotation of the coupler.

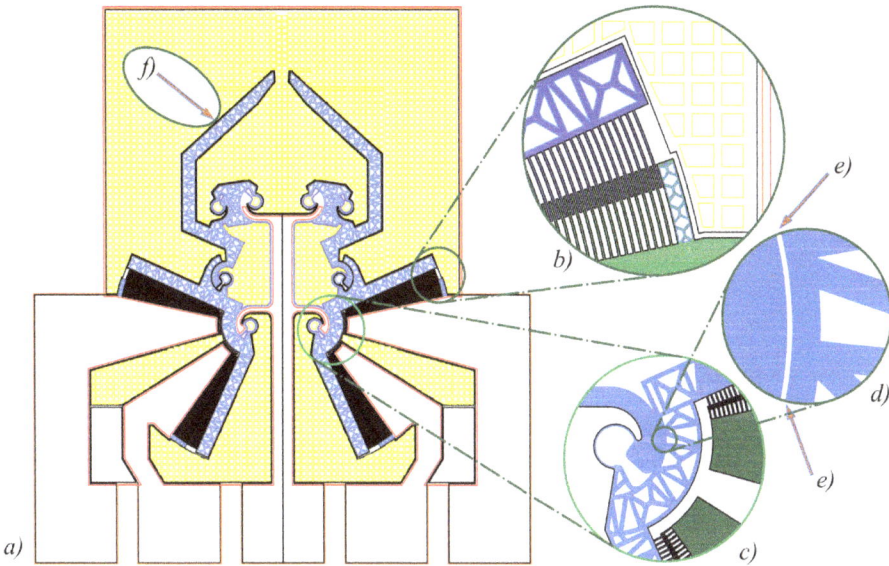

Figure 1. The mask adopted during Deep Reactive-Ion Etching (DRIE) process (**a**); details of the fingers of the comb drive (**b**); the Conjugate Surface Flexure Hinge (CSFH) hinge (**c**); the gap between the conjugate surfaces (**e**) and the left-hand side jaw (**f**).

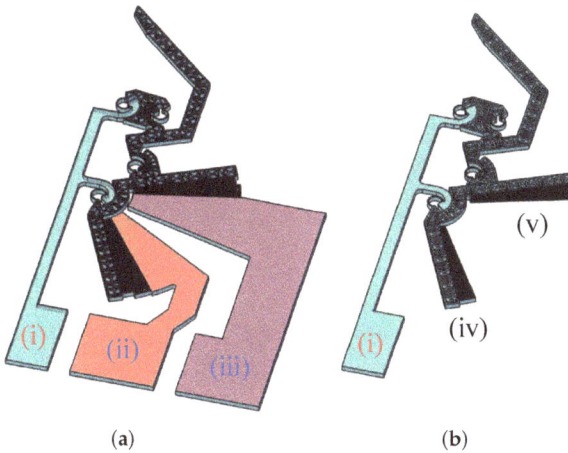

Figure 2. Model of the right jaw of the microgripper with (**a**) and without (**b**) the control pads: (i) device anchor; (ii) opening control pad; (iii) closing control pad; (iv) and (v) mobile fingers engaged with (ii) and (iii), respectively.

3. Fabrication

The fabrication stage has been performed starting from a 6-inch SEMI standard [48] silicon on insulator (SOI) wafer. Boron has been used as dopant for the wafer SOI device layer, with a resultant resistivity equal to 2–4 Ω·cm. Each wafer has a 500-μm handle layer that works as the device support. A 40 μm thick silicon layer works as device layer and is bonded at the top of the wafer. The device (top) layer is the most important one because the suspended moving subparts are therein patterned.

Between the support and the top layer, a 2-µm oxide layer stops etching during silicon patterning. This layer is also necessary to support the top layer before the release of the devices. The support and top layers of the wafer need to be both patterned and so Bosch deep reaction-ion etching (DRIE) process is applied to etch silicon down to the buried oxide layer. The DRIE was performed with an Alcatel AMS200 ICP plasma etcher (Alcatel-Lucent, Paris, France), using a standard two-step bosh recipe with fluorine chemistry. The initial step consists in the deposition of a multilayer mask to provide masking for DRIE etching process on both sides. A 150 nm silicon dioxide is firstly deposited. Then a 200 nm aluminum film is sputtered. A 100 nm film titanium is finally sputtered and standard photolithography is used to pattern this layer stack. The front side is exposed by means of an i-line stepper, while the backside is aligned by means of a broadband mask aligner. Finally, stack etching takes place in IC standard plasma reactors.

After masks patterning, an Alcatel SMS200 (Alcatel-Lucent, Paris, France), etcher is used to etch silicon through DRIE Bosch process. Front and backside etching deepness is equal to 40 µm and 500 µm, respectively.

At the end of this process, the front and back geometry is transferred on the wafer sides. Therefore, the silicon dioxide intermediate layer must be removed by wet etching in hydrofluoric acid based solution. This treatment allows the device to be released and freely move. During wet etching, residual DRIE mask layers are etched as well.

For the sake of electrostatic actuation, some electrical connections are developed by using the physical vapor deposition of an aluminum layer on the front side.

Aluminum hard mask allows the process to greatly reduce the mask layer thickness. This simplifies sub-micron features patterning, eliminating the obstacle of high aspect ratio thick mask etching. Further, aluminum grants an excellent feature size control during pattern transfer from the mask to silicon, as it eliminates mask edge erosion. The three layer structure and the whole fabrication process have been fully described by Bagolini et al. [20]. In the micro-gripper device, an aspect ratio up to 20 is implemented, but the fabrication module is developed for higher aspect ratio which is not part of present work.

4. Contact and Mobility Tests

As mentioned in Section 2, the adopted mechanism is composed of two symmetric four-bar linkages, each one having four CSFH compliant hinges. Since each of the two mobile suspended structures is held in place by only two 20×5 µm^2 cross sectional areas, some concern has arisen about the microsystem capability to bear externally applied loads. Hence, the purpose of the present paper consists in testing the mechanical structure of the developed microgripper. For this reason, the experimental activities have been arranged in such a way to provide information on the resistance of the structure, whereas its operational capability in unloaded conditions have been previously investigated [28,49].

In the present paper, a contact force has been externally applied to the right arm of the microgripper. This external load is much more invasive than the torque which is exerted by the electrostatic actuation on the crank-link, because it transmits to the block of mobile fingers not only a moment, but also a force. Such force has generally a radial component, the latter being much dangerous for the microsystem, because it pushes mobile fingers against the fixed ones.

The microgripper under analysis has been tested by means of an Agilent probestation equipped with binocular microscope and needle probes with tri-axial micrometer positioning control. The kinematic functionality of the device has been verified by monitoring the microgripper response to a force exerted by a probe through the contact area.

A series of mechanical tests have been conducted by using the three-axis micro probe (Suss PH 150 probehead). As reported in Figure 3, the probe was moved in such a way to mimic the electrostatic action exerted by the comb-drive, rotating the floating part toward the corresponding anchor.

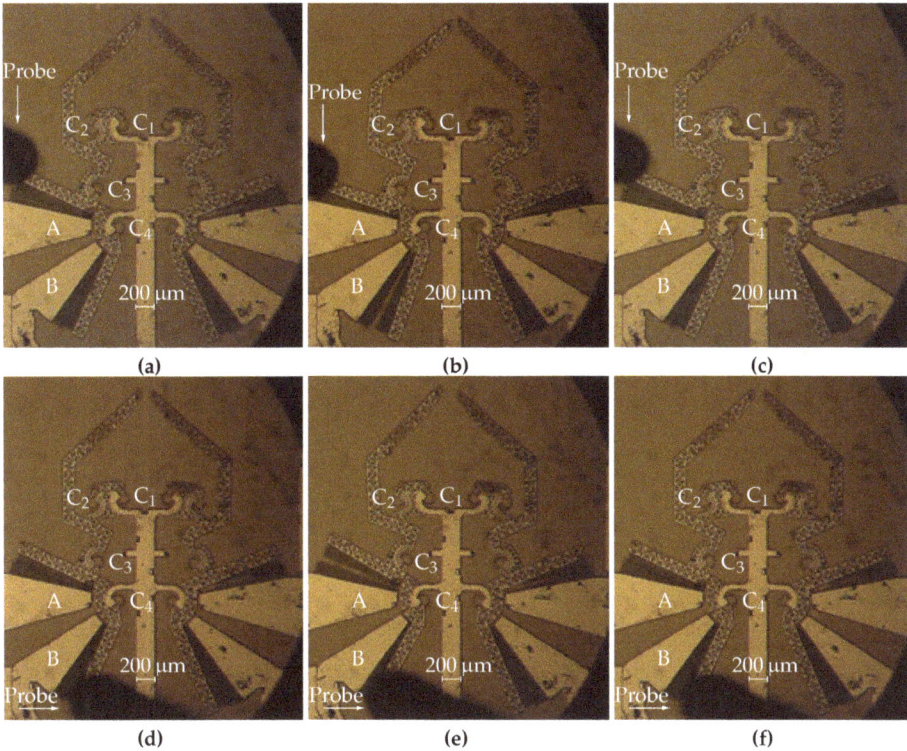

Figure 3. Steps of the mechanical testing procedure: (**a**) Probe positioning near the closing comb-drive (A); (**b**) Close configuration: stroke limit for the closing comb-drive (A); (**c**) Neutral configuration: releasing of the closing comb-drive (A); (**d**) Probe positioning near the opening comb-drive (B); (**e**) Open configuration: stroke limit for the opening comb-drive (B); (**f**) Neutral configuration: releasing of the opening comb-drive (B).

The test is performed considering the following steps:

(a) the probe is positioned in proximity of the closing comb-drive: before contact, the compliant mechanism stands in neutral configuration;

(b) the probe contacts the device and gently pushes the floating part until the maximum rotation is reached. Therefore, the gripper jaw follows a closing trajectory until the device achieve the close configuration (Figure 3b). During this stage, while the rotation angle between floating and anchored parts of the closing comb-drive decreases, the relative rotation between the same parts of the opening comb-drive increases;

(c) the probe is brought back to the initial position; in this phase, the comb-drive is gradually released and the gripper jaw follows an opening trajectory, until the microgripper achieves again its neutral configuration (Figure 3c);

(d) the probe is positioned in proximity of the opening comb-drive: as in the previous case, before contact, the microgripper is in neutral configuration (Figure 3d);

(e) the probe contacts and pushes the floating part to the comb-drive limit position: the gripper jaw follows an opening trajectory until the device achieve the open configuration (Figure 3e). While the rotation between the floating and anchored parts of the opening comb-drive decreases, the relative rotation between the same parts of the closing comb-drive increases (Figure 3e);

(f) the probe is repositioned at the starting point Figure 3d: the comb-drive is released and the gripper jaw follows an closing trajectory, until the neutral configuration is achieved again (Figure 3f).

Figure 4 shows three overlapping images reporting the left arm of the microgripper in close, neutral, and open configurations. The probe force could not be directly measured, due to the particular adopted experimental set. However, the applied direction could be obtained by recording the tip displacements from the neutral position, while the force magnitude was measured by means of FEA. In fact, a magnitude of the applied force equal to 320 µN was calculated as compatible with the observed close and open configurations. Figure 4 presents also a schematic representation of the two forces corresponding to the close and opening maximal configurations.

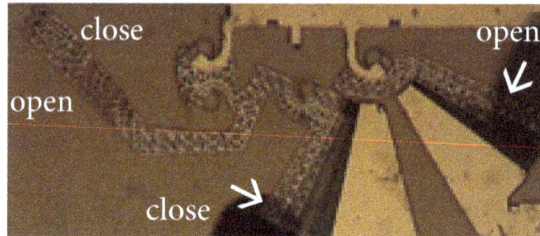

Figure 4. Overlapping images of the microgripper in close, neutral, and open configurations.

Contact Analysis on the Conjugate Surfaces

The CSFH is able to modify substantially the static and dynamic behaviors of the whole microsystem which it belongs to, depending on the occurrence of the contact between the conjugate surfaces. In fact, once the two conjugate surfaces get in touch, a reaction force arises between them. Actually, the purpose of CSFH hinges is to use this reaction force in order to restrict the displacements of the centers of the relative rotations between adjacent links within the size of the gap between the conjugate surface (actually about 5 µm). This makes the microsystem quite stable. However, contacts within a CSFH hinge could be activated too late, namely, after the breaking of the curved beam, which represent the elastic part of the CSFH. Therefore, experimental tests of the system response to externally applied forces are fundamental to validate the FEA model and the structural design of the whole system. The conditions under which this contact occurs are useful to optimize the final layout.

In order to analyze different contact conditions, a force $\mathbf{F}$, illustrated in Figure 5, is introduced to represent the action that could be applied to the tip during the gripping task or the positioning phase. The right hand side four-bar linkage is also illustrated together with the CSFH's C_1, C_2, C_3 and C_4. The force magnitude $|\mathbf{F}|$ and orientation φ have been adopted as variable parameters, with the purpose of detecting their critical values in correspondence of which the conjugate profiles of the four embedded CSFH get in touch.

Six different values of the magnitude $|\mathbf{F}|$ have been applied, with the assumption that $\mathbf{F}$ acts on the device working plane, with $F_z = 0$. The force direction is identified in the plane by means of the angle φ between $\mathbf{F}$ and the x axis. The full span from $0°$ and $360°$ has been investigated for φ. It is clear, from the figure, that $\mathbf{F}$ induces deformations that are concentrated specially on the curved beams, which form the elastic part of the CSFHs. These flexure elements deflect under the action of the internal loads and therefore the conjugate surfaces may get in contact.

A contact chart has been obtained for each CSFH C_i, as represented in Figure 6, by iterating FEA for discrete number of values of the angle $\varphi = 0, \ldots, 2\pi$ and six values of the force magnitude. Each chart allows designer to immediately understand which are the CSFHs which present contact for the given pair of parameters. For example, contact occurrence in C_i, $i = 1, \ldots, 4$ is highlighted by red colored arches and the red sectors are those corresponding to the directions and magnitudes of $\mathbf{F}$ which induce contact in C_i hinge.

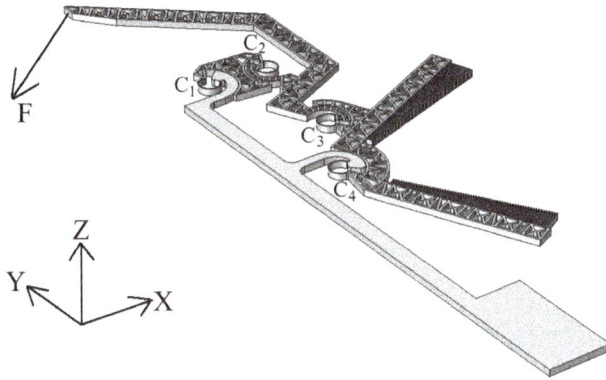

Figure 5. Static load to modify the CSFH contact.

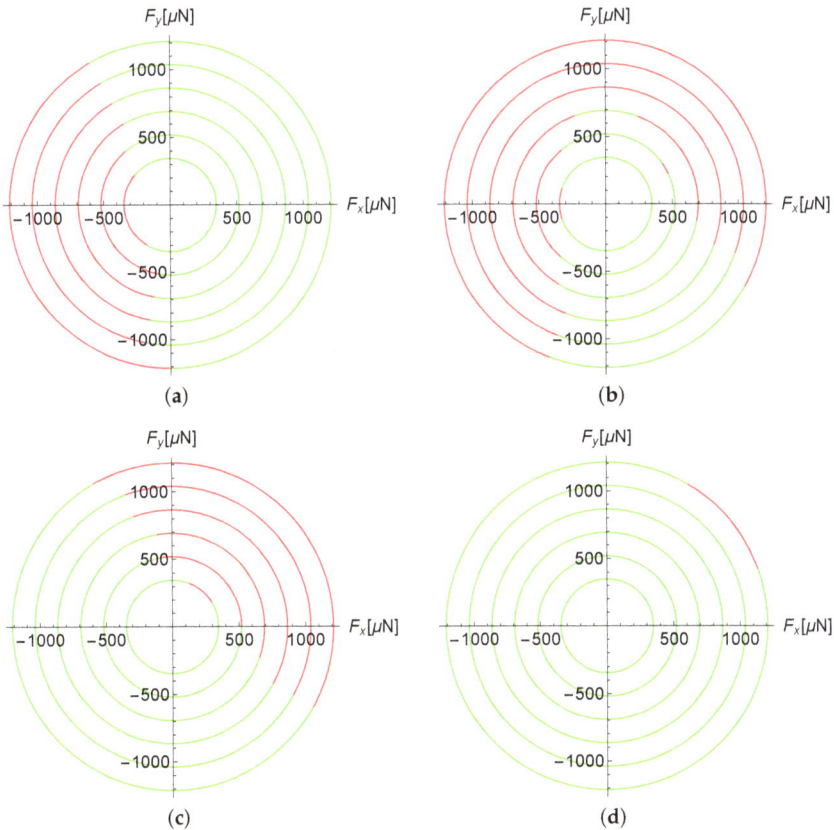

Figure 6. Force vs. contact diagrams: contact charts for the joint C_1 (**a**); C_2 (**b**); C_3 (**c**); and C_4 (**d**), respectively.

It is possible to observe that there are different conjugate surfaces of the CSFHs into contact for different angles. C_2 and C_4 are the CSFHs where the conjugate profiles more or less easily get in contact, respectively. Figure 7 shows, in particular, the various CSFHs configurations for different angles φ. The magnitude of the adopted force $|F|$ (some hundreds of µN) is comparable to the value of the forces applied to the comb drive mobile finger set (320 µN, as depicted in Figure 4) to fully close or open the configuration.

Figure 7. Configurations of CSFHs with $|\mathbf{F}| = 1 \times 10^{-5}$ (N): (**a**) C_1 and C_2 at $\varphi = 50°$; (**b**) C_3 and C_4 at $\varphi = 50°$; (**c**) C_1 and C_2 at $\varphi = 180°$; (**d**) C_3 and C_4 at $\varphi = 180°$; (**e**) C_1 and C_2 at $\varphi = 280°$; (**f**) C_3 and C_4 at $\varphi = 280°$.

5. Vibrational Mode Analysis

The versatility of the CSFH makes them useful in many applications such as micro mechanics, biology, etc. However, in many of these fields the analysis of the vibrations is essential on both negative and positive effects. In fact, the induced vibrations can lead to significant positioning errors [50,51] or undesirable contacts between the fingers. Likewise, jaws vibrations can be exploited to release micro objects [52] or to analyse the mechanical characteristics of the soft biological tissues to study possible pathologies [53].

Direct experimental acquisition of the vibrational modes is not easy for MEMS, although there are some successful cases. For example, in-plane low-frequency vibration was measured by means of circular grating Talbot interferometer [54]. With this method vibration from 1 to 25 Hz has been monitored in a plate system. Scanning laser Doppler vibrometry and experimental modal analysis have been also successfully applied to a pair of micro-cantilevers [55]. A charge-coupled device camera and synchronized pulsating illumination has been also used to measure sub-micrometer in-plane dynamics of MEMS devices with nanoscale precision [56]. Another interesting case of optical detection consists in the acquisition of the electromechanical response of a MEMS-technology based micro-mirror used in scanning pico-projectors [57].

In this paper, the experimental activities have been restricted to the static force analysis only and therefore the vibrational modes have been conveniently obtained via Finite Element Analysis. The modal analysis has been performed with the ANSYS © (Canonsburg, PA, USA) software [58]. The first six, eigenmodes and eigenfrequencies have been reported in Figure 8. In this analysis the device anchor has been considered fixed.

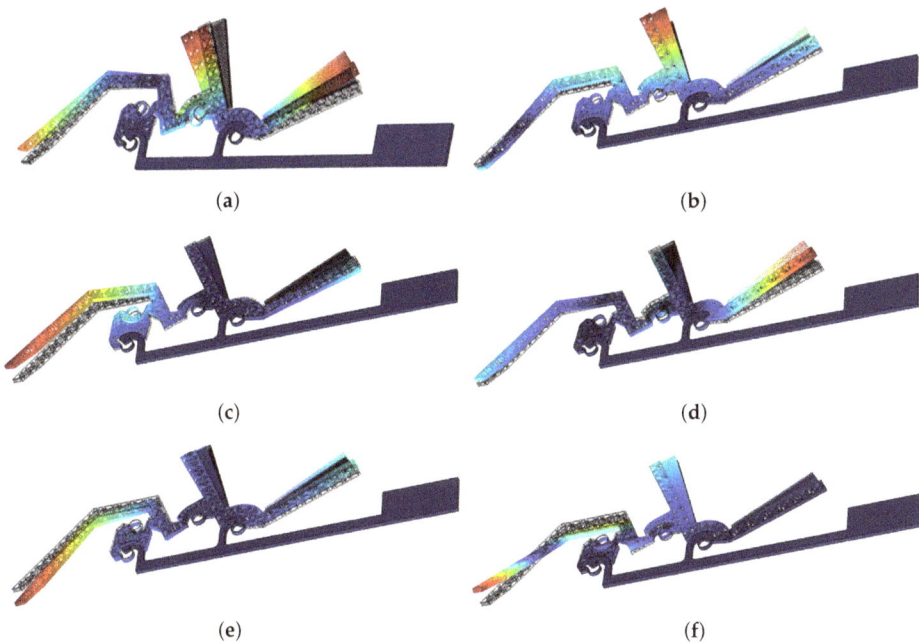

(a)   (b)

(c)   (d)

(e)   (f)

Figure 8. Eigenmodes and eigenfrequencies: first ($f_1 = 1.5147$ kHz) (**a**); second ($f_2 = 1.9433$ kHz) (**b**); third ($f_3 = 2.4749$ kHz) (**c**); fourth ($f_4 = 2.9982$ kHz) (**d**) ; fifth ($f_5 = 6.4103$ kHz) (**e**); and sixth ($f_6 = 6.4618$ kHz) (**f**) eigenmode, respectively.

The relative displacements between the anchored and the rotating fingers are tolerable only if they correspond to a relative rotation where the center is coincident with the center of the conjugate profiles. Therefore, it is very important to understand the nature of the relative motion for all the possible vibration modes. In particular, the first, second and fourth modes present radial displacements for the fingers and therefore finger contact appears to be theoretically possible.

6. Results and Discussions

The results obtained by means of FEA can be displayed, for each vibrational mode, as animated sequences of intermediate deformed configurations. This opportunity has been taken to identify, for each mode, those elements with the minimum displacements during the vibration motion, when a certain mode is excited. Considering also the numerical data, these elements can be easily identified for the first 6 modes and the behavior of the structure can be physically interpreted by introducing nodal oscillation axes.

When the system is excited at the first mode (i), three subparts of the structures behave as pseudo-rigid bodies that correspond to the two rockers and the coupler links. Motion is provided by the four flexure hinges and therefore a relative rotation axis appears for each hinge. During this motion all the subparts move within the mask plane and so the structure is characterized by an in-plane motion. The relative rotation axes are orthogonal to the plane and their intersections with the plane are represented in Figure 9a. Hence, first natural mode consists of a motion that is coincident with the motion for which the system has been designed. As a consequence, this kind of deformation is compatible with the geometry of the fingers because the relative rotation axes are practically coincident with the CSFH rotation axes.

The structure oscillations related to the second mode (ii) reveal that the whole structure behaves approximately as if it was a whole plate which rotates around the axis passing through the centers of the framed CSFH, as depicted in Figure 9b. This axis belongs to the main plane and therefore the oscillation will take place out of the plane. However, the mobile fingers are completely positioned by one side of the axis and so their motion goes along a direction which is orthogonal to the gap between the fingers. This means that finger contact remains quite unlikely.

Considering the third mode (iii), a nodal axis through the centers of two adjacent hinges has been identified and illustrated, as the axis (iii), in Figure 9b. In case the system is excited with the third mode, it will behave, approximately, as it was composed by a flexible plate that oscillates around the axis (iii) with additional deformations due to the presence of the anchored parts. Once again, axis (iii) belongs to the main plane and therefore the fingers will be affected also by an out-plane motion, with limited effect on reciprocal contact likelihood.

Taking into account the fourth mode (iv), the modal analysis shows that the microgripper inflects around two the parallel axes (iv)-a and (iv)-b depicted in Figure 9c. These axes belong to the plane and so the displacements will be out-planar directed. The system is roughly comparable to a flexible plate oscillating around nodal axes (iv)-a and (iv)-b, with three different zones. For example, while the central zone is up, the lateral parts will be down, and vice versa.

In view of preventing comb drives from finger contact, the fifth node of vibration presents the most difficult circumstance under which the microgripper behaves, approximately, as a pseudo-rigid plate which rotates around a point that is positioned within the internal area of the four-bar linkage, as reported in Figure 9d. This is possible because the CSFHs behave as suspension springs. Unfortunately, this motion is rather dangerous for the comb drive fingers, because the mobile finger sets do not rotate about the CSFH centers. This means that the curved fingers follow no more the natural span of the fixed gaps and so they collide with the fixed sets of fingers. This circumstance is depicted in Figure 10 which shows that the mobile finger set has a radial component of the displacement, having lost the original rotation center, with ineluctable mechanical interference.

The vibrational shape related to the sixth mode is rather complicated to be described. In fact, three nodal axes can be identified (Figure 9e). These axes are all in the fabrication plane. For this case, the microgripper behaves as a flexible plate which inflects around the nodal lines. Since the axes are in the plane, out-plane displacements are the most significant and so no great problem is expected for the fingers.

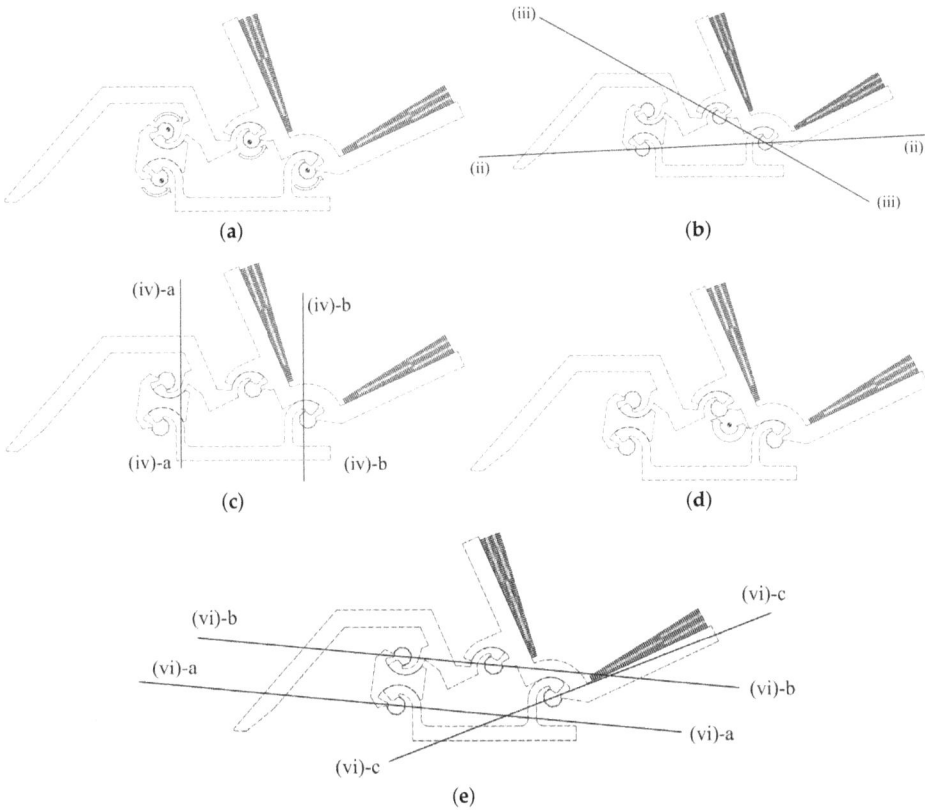

Figure 9. Pseudo-axes of rotation corresponding to the first 6 vibrational modes: (**a**) In-plane rotation axes for mode (i); (**b**) Out-plane rotation axes for modes (ii) and (iii); (**c**) Out-plane rotation axis for mode (iv); (**d**) In-plane rotation axis for mode (v); (**e**) Out-plane rotation axis for mode (vi).

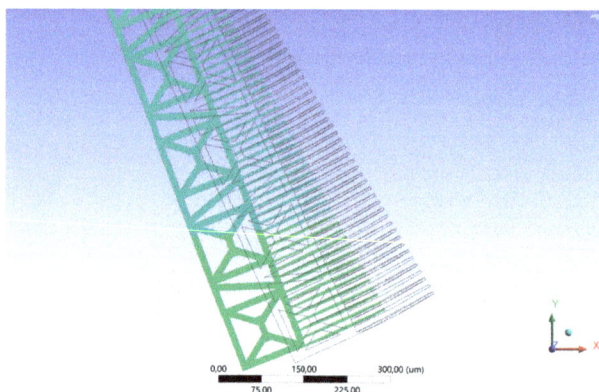

Figure 10. Critical configuration of the comb drive fingers when mode (v) is excited.

Considering that the amplitude of the dumped response to external excitation generally have higher values at the lowest frequencies (see for example Reference [59]), the first six modes only have been considered. Furthermore, given the obtained values of the first six natural frequencies, listed in Table 1, the fifth and the sixth modes will have limited influence with respect to the previous four, because their values is more than twice the fourth frequency. Since fingers contact is expected only at the fifth critical mode, its occurrence likelihood is rather limited.

Table 1. Critical modes for comb drives fingers.

Frequency	kHz	Critical	Plane	Rotation Axes
i	1.51	no	in	4
ii	1.94	no	out	1
iii	2.47	no	out	1
iv	3.00	no	out	2
v	6.41	yes	in	1
vi	6.46	no	out	3

7. Conclusions

The present investigation has shown that the microgripper under analysis it is able to sustain large modifications of the configuration toward both the opening and the closing positions. Considering the size of the embedded flexure hinges, which consist of silicon curved beams with a $5 \times 40 \ \mu m^2$ cross sectional area, an evident structural robustness has been experimentally observed. In fact, the microgripper has been able to resist to all the actions exerted by the probe. The second part of the investigation has been based on a numerical approach. Firstly, some design charts have been obtained to optimize the likelihood of the contact between the conjugate profiles. Secondly, FEA has been used to detect the main vibrational modes, whose analysis is necessary to prevent fingers contacts. The whole investigation has therefore confirmed that the microgripper under analysis is robust in operational conditions and promising for the applications.

Author Contributions: Conceptualization, all the authors; Methodology, F.B. and M.V.; Software, F.B.; Validation, all the authors; Formal analysis, F.B. and M.V.; Investigation and resources for fabrication, A.B. and P.B.; Data Curation, all the authors; Writing—original draft preparation, Writing—review & Editing & Visualization, M.V. and N.P.B.; Supervision, P.B. and N.P.B.

Funding: This research received no external funding.

Conflicts of Interest: The authors declare no conflict of interest.

References

1. Lobontiu, N.; Garcia, E.; Canfield, S. Torsional stiffness of several variable rectangular cross-section flexure hinges for macro-scale and MEMS applications. *Smart Mater. Struct.* **2004**, *13*, 12–19. [CrossRef]
2. Gad-el Hak, M.E. (Ed.) *The MEMS Handbook*, 2nd ed.; Mechanical and Aerospace Engineering Series, 3 Volume Set; CRC Press: Boca Raton, FL, USA, 2005.
3. Bhushan, B.E. (Ed.) *Springer Handbook of Nanotechnology*; Springer Handbooks; Springer: Berlin, Germany, 2017.
4. Dochshanov, A.; Verotti, M.; Belfiore, N.P. A Comprehensive Survey on Microgrippers Design: Operational Strategy. *J. Mech. Des. Trans. ASME* **2017**, *139*, 070801. [CrossRef]
5. Verotti, M.; Dochshanov, A.; Belfiore, N.P. A Comprehensive Survey on Microgrippers Design: Mechanical Structure. *J. Mech. Des. Trans. ASME* **2017**, *139*, 060801. [CrossRef]
6. Cecil, J.; Powell, D.; Vasquez, D. Assembly and manipulation of micro devices—A state of the art survey. *Robot. Comput.-Integr. Manuf.* **2007**, *23*, 580–588. [CrossRef]
7. Zhang, J.; Lu, K.; Chen, W.; Jiang, J.; Chen, W. Monolithically integrated two-axis microgripper for polarization maintaining in optical fiber assembly. *Rev. Sci. Instrum.* **2015**, *86*, 025105. [CrossRef] [PubMed]
8. Chen, W.; Shi, X.; Chen, W.; Zhang, J. A two degree of freedom micro-gripper with grasping and rotating functions for optical fibers assembling. *Rev. Sci. Instrum.* **2013**, *84*, 115111. [CrossRef] [PubMed]
9. Ingber, D.E. Cellular mechanotransduction: Putting all the pieces together again. *FASEB J.* **2006**, *20*, 811–827. [CrossRef] [PubMed]
10. Moeendarbary, E.; Harris, A.R. Cell mechanics: Principles, practices, and prospects. *Wiley Interdiscip. Rev.* **2014**, *6*, 371–388. [CrossRef]
11. Beyeler, F.; Neild, A.; Oberti, S.; Bell, D.J.; Sun, Y.; Dual, J.; Nelson, B.J. Monolithically fabricated microgripper with integrated force sensor for manipulating microobjects and biological cells aligned in an ultrasonic field. *J. Microelectromech. Syst.* **2007**, *16*, 7–15. [CrossRef]
12. Kim, K.; Liu, X.; Zhang, Y.; Sun, Y. Nanonewton force-controlled manipulation of biological cells using a monolithic MEMS microgripper with two-axis force feedback. *J. Micromech. Microeng.* **2008**, *18*, 055013. [CrossRef]
13. Di Giamberardino, P.; Bagolini, A.; Bellutti, P.; Rudas, I.; Verotti, M.; Botta, F.; Belfiore, N.P. New MEMS tweezers for the viscoelastic characterization of soft materials at the microscale. *Micromachines* **2017**, *9*, 15. [CrossRef] [PubMed]
14. Di Giamberardino, P.; Aceto, M.; Giannini, O.; Verotti, M. Recursive Least Squares Filtering Algorithms for On-Line Viscoelastic Characterization of Biosamples. *Actuators* **2018**, *7*, 74. [CrossRef]
15. Lan, C.C.; Lin, C.M.; Fan, C.H. A self-sensing microgripper module with wide handling ranges. *IEEE/ASME Trans. Mechatron.* **2011**, *16*, 141–150, doi:10.1109/TMECH.2009.2037495. [CrossRef]
16. Shen, Y.; Winder, E.; Xi, N.; Pomeroy, C.; Wejinya, U. Closed-loop optimal control-enabled piezoelectric microforce sensors. *IEEE/ASME Trans. Mechatron.* **2006**, *11*, 420–427, doi:10.1109/TMECH.2006.878555. [CrossRef]
17. Lu, Z.; Chen, P.; Ganapathy, A.; Zhao, G.; Nam, J.; Yang, G.; Burdet, E.; Teo, C.; Meng, Q.; Lin, W. A force-feedback control system for micro-assembly. *J. Micromech. Microeng.* **2006**, *16*, 1861–1868, doi:10.1088/0960-1317/16/9/015. [CrossRef]
18. Rakotondrabe, M.; Rabenorosoa, K.; Agnus, J.; Chaillet, N. Robust feedforward-feedback control of a nonlinear and oscillating 2-DoF piezocantilever. *IEEE Trans. Autom. Sci. Eng.* **2011**, *8*, 506–519, doi:10.1109/TASE.2010.2099218. [CrossRef]
19. Belfiore, N.P.; Broggiato, G.B.; Verotti, M.; Balucani, M.; Crescenzi, R.; Bagolini, A.; Bellutti, P.; Boscardin, M. Simulation and construction of a mems CSFH based microgripper. *Int. J. Mech. Control* **2015**, *16*, 21–30.
20. Bagolini, A.; Ronchin, S.; Bellutti, P.; Chistè, M.; Verotti, M.; Belfiore, N.P. Fabrication of Novel MEMS Microgrippers by Deep Reactive Ion Etching With Metal Hard Mask. *J. Microelectromech. Syst.* **2017**, *26*, 926–934. [CrossRef]
21. Belfiore, N.P.; Scaccia, M.; Ianniello, F.; Presta, M. Selective Compliance Hinge. U.S. Patent 8,191,204 B2, 5 June 2012.
22. Belfiore, N.; Simeone, P. Inverse kinetostatic analysis of compliant four-bar linkages. *Mech. Mach. Theory* **2013**, *69*, 350–372. [CrossRef]

23. Nenzi, P.; Crescenzi, R.; Dolgyi, A.; Klyshko, A.; Bondarenko, V.; Belfiore, N.; Balucani, M. High density compliant contacting technology for integrated high power modules in automotive applications. In Proceedings of the Electronic Components and Technology Conference, San Diego, CA, USA, 29 May–1 June 2012; pp. 1976–1983.

24. Balucani, M.; Belfiore, N.; Crescenzi, R.; Genua, M.; Verotti, M. Developing and modeling a plane 3 DOF compliant micromanipulator by means of a dedicated MBS code. In Proceedings of the 2011 NSTI Nanotechnology Conference and Expo (NSTI-Nanotech 2011), Boston, MA, USA, 13–16 June 2011; Volume 2, pp. 659–662.

25. Balucani, M.; Belfiore, N.; Crescenzi, R.; Verotti, M. The development of a MEMS/NEMS-based 3 D.O.F. compliant micro robot. *Int. J. Mech. Control* **2011**, *12*, 3–10.

26. Belfiore, N.; Balucani, M.; Crescenzi, R.; Verotti, M. Performance analysis of compliant MEMS parallel robots through pseudo-rigid-body model synthesis. In Proceedings of the ASME 2012 11th Biennial Conference on Engineering Systems Design and Analysis (ESDA 2012), Nantes, France, 2–4 July 2012; Volume 3, pp. 329–334.

27. Belfiore, N.P.; Prosperi, G.; Crescenzi, R. A simple application of conjugate profile theory to the development of a silicon micro tribometer. In Proceedings of the ASME 2014 12th Biennial Conference on Engineering Systems Design and Analysis (ESDA 2014), Copenhagen, Denmark, 25–27 July 2014; Web Portal ASME (American Society of Mechanical Engineers): New York, NY, USA, 2014; Volume 2.

28. Crescenzi, R.; Balucani, M.; Belfiore, N.P. Operational characterization of CSFH MEMS technology based hinges. *J. Micromech. Microeng.* **2018**, *28*, 055012. [CrossRef]

29. Zeman, M.J.F.; Bordatchev, E.V.; Knopf, G.K. Design, kinematic modeling and performance testing of an electro-thermally driven microgripper for micromanipulation applications. *J. Micromech. Microeng.* **2006**, *16*, 1540. [CrossRef]

30. Zhang, R.; Chu, J.; Wang, H.; Chen, Z. A multipurpose electrothermal microgripper for biological micro-manipulation. *Microsyst. Technol.* **2013**, *19*, 89–97, doi:10.1007/s00542-012-1567-0. [CrossRef]

31. Chang, R.; Wang, H.; Wang, Y. Development of mesoscopic polymer gripper system guided by precision design axioms. *Precis. Eng.* **2003**, *27*, 362–369, doi:10.1016/S0141-6359(03)00042-4. [CrossRef]

32. Kohl, M.; Just, E.; Pfleging, W.; Miyazaki, S. SMA microgripper with integrated antagonism. *Sens. Actuators A Phys.* **2000**, *83*, 208–213. [CrossRef]

33. Chen, T.; Chen, L.; Sun, L.; Wang, J.; Li, X. A sidewall piezoresistive force sensor used in a MEMS gripper. In Proceedings of the International Conference on Intelligent Robotics and Applications, Wuhan, China, 15–17 October 2008; Springer: Berlin, Germany, 2008; pp. 207–216.

34. Zubir, M.N.M.; Shirinzadeh, B.; Tian, Y. Development of a novel flexure-based microgripper for high precision micro-object manipulation. *Sens. Actuators A Phys.* **2009**, *150*, 257–266, doi:10.1016/j.sna.2009.01.016. [CrossRef]

35. Dong, J.; Ferreira, P.M. Electrostatically actuated cantilever with SOI-MEMS parallel kinematic XY stage. *J. Microelectromech. Syst.* **2009**, *18*, 641–651. [CrossRef]

36. Selvakumar, A.; Najafi, K. Vertical comb array microactuators. *J. Microelectromech. Syst.* **2003**, *12*, 440–449. [CrossRef]

37. Xie, H.; Fedder, G.K. Vertical comb-finger capacitive actuation and sensing for CMOS-MEMS. *Sens. Actuators A Phys.* **2002**, *95*, 212–221. [CrossRef]

38. Chen, T.; Sun, L.; Chen, L.; Rong, W.; Li, X. A hybrid-type electrostatically driven microgripper with an integrated vacuum tool. *Sens. Actuators A Phys.* **2010**, *158*, 320–327, doi:10.1016/j.sna.2010.01.001. [CrossRef]

39. Wierzbicki, R.; Houston, K.; Heerlein, H.; Barth, W.; Debski, T.; Eisinberg, A.; Menciassi, A.; Carrozza, M.; Dario, P. Design and fabrication of an electrostatically driven microgripper for blood vessel manipulation. *Microelectron. Eng.* **2006**, *83*, 1651–1654, doi:10.1016/j.mee.2006.01.110. [CrossRef]

40. Hou, M.T.K.; Huang, J.Y.; Jiang, S.S.; Yeh, J.A. In-plane rotary comb-drive actuator for a variable optical attenuator. *J. Micro/Nanolithogr. MEMS MOEMS* **2008**, *7*, 043015.

41. Yeh, J.; Jiang, S.S.; Lee, C. MOEMS variable optical attenuators using rotary comb drive actuators. *IEEE Photonics Technol. Lett.* **2006**, *18*, 1170–1172, doi:10.1109/LPT.2006.873959. [CrossRef]

42. Cecchi, R.; Verotti, M.; Capata, R.; Dochshanov, A.; Broggiato, G.; Crescenzi, R.; Balucani, M.; Natali, S.; Razzano, G.; Lucchese, F.; et al. Development of micro-grippers for tissue and cell manipulation with direct morphological comparison. *Micromachines* **2015**, *6*, 1710–1728, doi:10.3390/mi6111451. [CrossRef]

43. Howell, L.L. Compliant Mechanisms. In *Encyclopedia of Nanotechnology*; Bhushan, B., Ed.; Springer: Dordrecht, The Netherlands, 2016; pp. 604–611.

44. Belfiore, N.P. Distributed Databases for the development of Mechanisms Topology. *Mech. Mach. Theory* **2000**, *35*, 1727–1744. [CrossRef]

45. Belfiore, N.P. Brief note on the concept of planarity for kinematic chains. *Mech. Mach. Theory* **2000**, *35*, 1745–1750. [CrossRef]

46. Sanò, P.; Verotti, M.; Bosetti, P.; Belfiore, N.P. Kinematic Synthesis of a D-Drive MEMS Device with Rigid-Body Replacement Method. *J. Mech. Des. Trans. ASME* **2018**, *140*, 075001. [CrossRef]

47. Verotti, M.; Dochshanov, A.; Belfiore, N.P. Compliance Synthesis of CSFH MEMS-Based Microgrippers. *J. Mech. Des. Trans. ASME* **2017**, *139*, 022301. [CrossRef]

48. SEMICON West. Beyond Smart. Available online: www.semifoundation.org (accessed on 9 October 2018).

49. Potrich, C.; Lunelli, L.; Bagolini, A.; Bellutti, P.; Pederzolli, C.; Verotti, M.; Belfiore, N.P. Innovative silicon microgrippers for biomedical applications: Design, mechanical simulation and evaluation of protein fouling. *Actuators* **2018**, *7*, 12. [CrossRef]

50. Demaghsi, H.; Mirzajani, H.; Ghavifekr, H.B. A novel electrostatic based microgripper (cellgripper) integrated with contact sensor and equipped with vibrating system to release particles actively. *Microsyst. Technol.* **2014**, *20*, 2191–2202. [CrossRef]

51. Park, J.; Moon, W. The systematic design and fabrication of a three-chopstick microgripper. *Int. J. Adv. Manuf. Technol.* **2005**, *26*, 251–261. [CrossRef]

52. Fang, Y.; Tan, X. A dynamic JKR model with application to vibrational release in micromanipulation. In Proceedings of the 2006 IEEE/RSJ International Conference on Intelligent Robots and Systems, Beijing, China, 9–15 October 2006; pp. 1341–1346.

53. Valtorta, D.; Mazza, E. Dynamic measurement of soft tissue viscoelastic properties with a torsional resonator device. *Med. Image Anal.* **2005**, *9*, 481–490. [CrossRef] [PubMed]

54. Agarwal, S.; Shakher, C. Low-frequency in-plane vibration monitoring/measurement using circular grating Talbot interferometer. *Opt. Eng.* **2018**, *57*, 054112. [CrossRef]

55. Rothberg, S.J.; Allen, M.S.; Castellini, P.; Di Maio, D.; Dirckx, J.J.J.; Ewins, D.J.; Halkon, B.J.; Muyshondt, P.; Paone, N.; Ryan, T.; et al. An international review of laser Doppler vibrometry: Making light work of vibration measurement. *Opt. Lasers Eng.* **2017**, *99*, 11–22. [CrossRef]

56. Warnat, S.; Forbrigger, C.; Kujath, M.; Hubbard, T. Nano-scale measurement of sub-micrometer MEMS in-plane dynamics using synchronized illumination. *J. Micromech. Microeng.* **2015**, *25*, 095004. [CrossRef]

57. Silva, G.; Carpignano, F.; Guerinoni, F.; Costantini, S.; De Fazio, M.; Merlo, S. Optical detection of the electromechanical response of MEMS micromirrors designed for scanning picoprojectors. *IEEE J. Sel. Top. Quantum Electron.* **2015**, *21*, 147–156. [CrossRef]

58. ANSYS, Inc., Canonsburg, PA, USA. Available online: https://www.ansys.com (accessed on 9 October 2018).

59. Weaver, W.; Timoshenko, S.P.; Young, D.H. *Vibration Problems in Engineering*; John Wiley and Sons: Hoboken, NJ, USA, 1990.

actuators

MDPI

Article

Recursive Least Squares Filtering Algorithms for On-Line Viscoelastic Characterization of Biosamples

Paolo Di Giamberardino [1], Maria Laura Aceto [1] and Oliviero Giannini [2] and Matteo Verotti [2,*]

[1] Department of Computer, Control and Management Engineering Antonio Ruberti, Sapienza University of Rome, Via Ariosto, 25, 00185 Rome, Italy; paolo.digiamberardino@uniroma1.it (P.D.G.); marialaura.aceto@uniroma1.it (M.L.A.)

[2] University of Rome Niccolò Cusano, Via Don Carlo Gnocchi, 3, 00166 Rome, Italy; oliviero.giannini@unicusano.it

* Correspondence: matteo.verotti@unicusano.it

Received: 31 August 2018; Accepted: 15 October 2018; Published: 22 October 2018

Abstract: The mechanical characterization of biological samples is a fundamental issue in biology and related fields, such as tissue and cell mechanics, regenerative medicine and diagnosis of diseases. In this paper, a novel approach for the identification of the stiffness and damping coefficients of biosamples is introduced. According to the proposed method, a MEMS-based microgripper in operational condition is used as a measurement tool. The mechanical model describing the dynamics of the gripper-sample system considers the pseudo-rigid body model for the microgripper, and the Kelvin–Voigt constitutive law of viscoelasticity for the sample. Then, two algorithms based on recursive least square (RLS) methods are implemented for the estimation of the mechanical coefficients, that are the forgetting factor based RLS and the normalised gradient based RLS algorithms. Numerical simulations are performed to verify the effectiveness of the proposed approach. Results confirm the feasibility of the method that enables the ability to perform simultaneously two tasks: sample manipulation and parameters identification.

Keywords: micromanipulation; microgripper; biological samples analysis; visco-elastic characteristic measurement; dynamic parameters estimation

1. Introduction

The mechanical characterization of biomaterials represents a crucial procedure in the fields of tissue engineering, tissue and cell mechanics, disease diagnosis and minimally invasive surgery (MIS) [1–3].

For instance, modeling the mechanical response of brain tissue can be useful in the understanding of traumatic brain injury [4] or blast–induced neurotrauma [5] mechanisms. In tissue engineering, scaffolds should achieve appropriate mechanical properties in order to ensure tissue growth. Therefore, before implantation, it is necessary to verify if these properties meet the specified requirements [6,7]. Many investigations have also been focused on the characterization and on the constitutive modeling of skin [8,9], cornea [10], vocal fold [11], skeletal muscle [12], and blood vessels [1] tissues. Regarding the circulatory system, it is worth mentioning that vascular stiffness is recognized as an important factor in diseases such as pulmonary arterial hypertension, kidney disease, and atherosclerosis [13]. Generally, the variation of tissue stiffness is a hallmark of several disease states, including fibrosis and some types of cancers [14]. Viscosity can also be considered as a biomarker for the metastatic potential of cancer cells [15,16], and the analysis of both elastic and viscous properties could be more effective in detecting and identifying specific diseases [17].

The characterization of the mechanical properties of biomaterials is important not only for in vitro analysis. For example, in MIS operations, tactile information is not available to the surgeon.

Therefore, the capability of on-line detecting and identifying tissues could play a fundamental role in the advancement of MIS procedures [3].

The mechanical properties of tissues have been investigated by using various methods, such as indentation and aspiration, and measuring deformations by means of ultrasound or magnetic resonance imaging techniques [18]. Surgical instruments have also been improved with sensing capabilities [19]. For example, a micro-tactile MEMS-based sensor to be integrated within MIS graspers was presented in Reference [3], whereas a sensor capable of determining the stiffness of biological tissues was modeled and tested in Reference [20].

At the cell level, experimental techniques for investigating mechanical properties include force-application techniques (e.g., micropipette aspiration, atomic force microscope probing, optical trapping), and force-sensing techniques (e.g., traction force microscopy, wrinkling membranes, micropost arrays). The technique selection depends on size of the biological sample, feature to be acquired, detrimental effects on the sample, spatial and force resolution, and accuracy [21,22].

Given the large number of experimental approaches, many mechanical models have been proposed in literature, based on the micro/nanostructural approach and on the continuum approach. The former model targets the sub-cell level, whereas the latter one treats the biological sample as a continuum material. Considering the continuum approach, many constitutive models have been proposed in literature, such as biphasic model, liquid drop models, solid viscoelastic models (Kelvin–Voigt, Maxwell, Zener) [23,24]. Once the mathematical model is available, a parameter estimation problem can be formulated and solved by means of different methods, such as least-squares [25], Kalman filtering [26], or inverse finite element [18,27] methods.

In the last decade, microgrippers became essential tools in the manipulation at the microscale [28–33], and can be adopted to develop novel techniques for the viscoelastic characterization of soft materials. In their previous investigation [34], the Authors proposed a method to evaluate the mechanical properties of a biomaterial sample based on the use of a MEMS microgripper. The mechanical model was developed considering the gripper–sample kinematics and the Kelvin–Voigt constitutive law of viscoelasticity, and a PID controller to determine stiffness and viscous parameters of the sample.

In this work, a parameter estimation problem is formulated starting from the model presented in Reference [34]. Recursive least squares filtering algorithms are implemented to characterize the mechanical properties of soft biomaterials, enabling the *on-line* identification of stiffness and viscosity coefficients. With respect to the other experimental techniques briefly mentioned above, that generally are implemented only for the mechanical testing procedure, the proposed approach enables the ability to perform the mechanical characterization *while* the sample is manipulated for a different purpose in another operation. Furthermore, this method does not require additional elements other than the measurement tool and the specimen as, for example, beads used with optical tweezers. Also, risks related to laser-induced damage and to large deformations of the sample (involved in optical traps and micropipette aspiration, respectively) are avoided.

The paper is organised as follows. The microdevice is introduced in Section 2, whereas the mathematical model is discussed in Section 3. The adopted identification algorithms are described in Section 4, and simulations and numerical results are reported and discussed in Section 5.

2. The Experimental Device

The microsystem considered in this paper consists of two silicon arms actuated by rotary comb drives, as shown in Figure 1. The electrostatic actuators exert the input torques that, through the conjugate surface flexure hinges (CSFH) deflections [35,36], move the arms from the initial position to perform the gripping task [37]. In the neutral configuration, the distance between the jaws is equal to 150 µm. The fabrication method relies on the application of deep reactive-ion etching on silicon-on-insulator wafers [38].

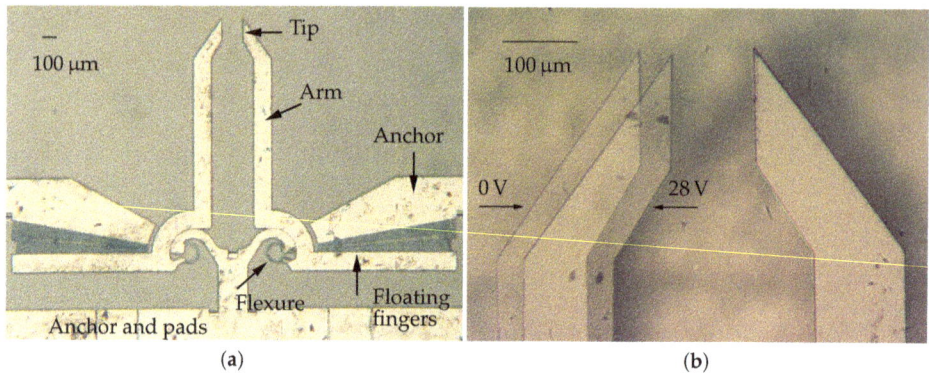

Figure 1. Optical microscope images of the device. The whole microgripper (**a**) and two overlapping frames showing the left arm in neutral (0 V) and actuated (28 V) configurations (**b**).

In their previous investigations [34,39], the authors estimated the mechanical characteristic of the sample considering input signals of suitable waveform. According to the proposed method, the sample is gripped with the left arm until the right arm reaches a predefined rotation, that serves as a reference signal for the implemented feedback control scheme. Therefore, the stiffness coefficient can be computed at steady state conditions [39]. In order to estimate also the sample viscosity, a small-amplitude sinusoidal signal was added to the left arm, and the viscous coefficient was obtained as a function of the input torque frequency [34].

In this paper, an estimation algorithm for the elastic and viscosity parameters of the gripped sample under generic control torques is presented. As demonstrated in the next Sections, the algorithm take advantage of the mathematical model structure that, despite its general non linearity, is linear with respect to the unknown parameters.

3. The Mathematical Model

In operative condition, the microdevice is gripping a sample between its jaws, as illustrated in Figure 2. More specifically, Figure 2a shows a schematic drawing of the compliant structure of the gripper. The contact points between jaws and sample are labelled as B and C, whereas the points A and D represent the centers of rotation of the left and right arms, respectively. Starting from the compliant mechanism, a pseudo-rigid body model (PRBM) can be obtained by substituting the constant-curvature CSFHs with revolute joints [35,40–42]. Figure 2b shows the PRBM of the microgripper: the quadrilateral $ABCD$ is a closed chain composed by the links AD (frame), AB and CD (arms), and BC (sample). All the links have a fixed length with the exception of BC, whose compression is at the basis of the measurement procedure. The proposed model takes into account the nonlinearity related to the kinematics of the gripper, but it assumes a linear behavior for stiffness coefficients k_2 and k_4. Moreover, the identification of the cell parameters inherently assumes that its dynamic behavior can be described by the Kelvin–Voight constitutive model of viscoelasticity.

With reference to Figure 2b, the following parameters can be introduced:

- l is the common length of the two links representing the arms, i.e., the distances AB and CD;
- d is the distance between the hinges (AD), i.e., the frame length;
- k_2, k_4 and k are the torsional stiffness of the two arms hinges and the stiffness of the sample, respectively;
- c_2, c_4 and c are the viscous damping coefficients of the two arms and of the sample, respectively;
- I_2 and I_4 are the moments of inertia of the left and right arms around A and D, respectively;
- τ_2 and τ_4 are the input torques generated by the left and right comb drives, respectively.

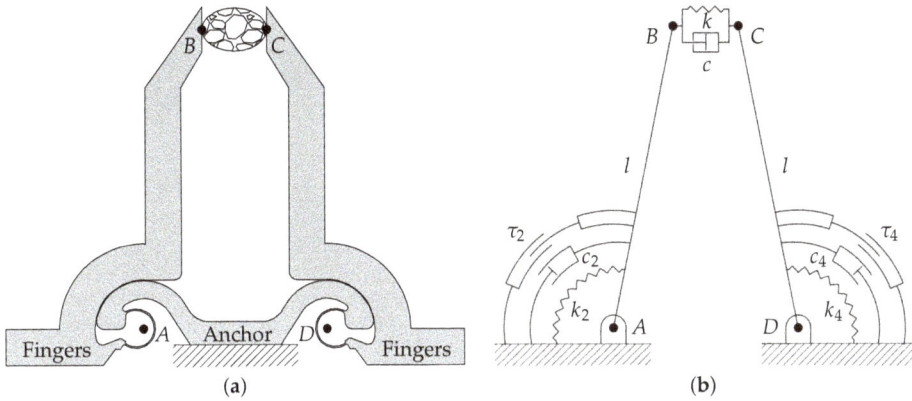

Figure 2. Schematic representations of the microgripper: compliant mechanism (**a**) and corresponding pseudo-rigid body model (**b**).

For the sake of simplicity in the model representation, all the considered variables are referred to the neutral layout: the gripper, in symmetrical configuration, is in contact with the sample but no deformation occurs. By following the notation introduced in Reference [34], and with reference to Figure 3a, the angles $\hat{\theta}_2$, $\hat{\theta}_3$, and $\hat{\theta}_4$ represent the reference orientations of the links AB, BC and DC, respectively, whereas $\hat{u}$ represents the reference length of the link BC. The notation $\tilde{\ }$ is used for the parameters in the deformed configuration, as shown in Figure 3b. Therefore, the angles $\theta_i = \tilde{\theta}_i - \hat{\theta}_i$ represent the relative angular displacements of the two links from their neutral configuration, with $i = 2$ for the left link and $i = 4$ for the right one. The orientation of BC follows the same notation, with $i = 3$ and $\hat{\theta}_3 = 0$, whereas the deformation is equal to $u = \tilde{u} - \hat{u}$. The values of the variables in the neutral configuration are given in Table 1.

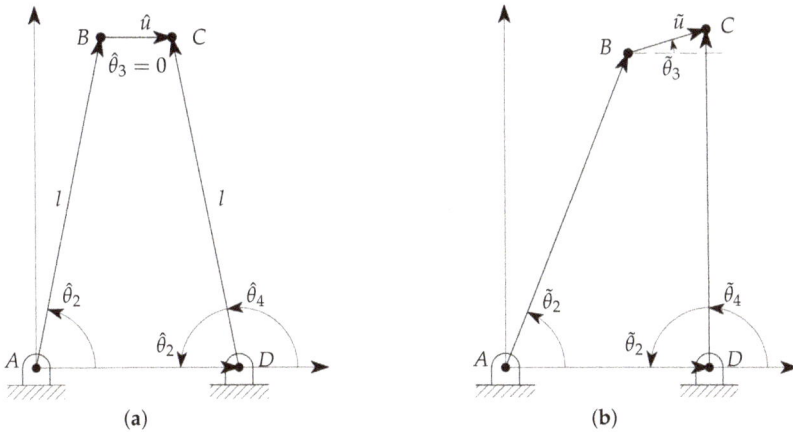

Figure 3. Nomenclature and model parameters in neutral (**a**) and general (**b**) configurations.

Table 1. Constants.

Parameter	Value	Unit
$\hat{\theta}_2$	1.44	rad
$\hat{\theta}_4$	1.70	rad
$\hat{u}$	150×10^{-6}	m

Assuming the inertia of the sample to be negligible, the dynamical model of the links can be written as

$$I_2\ddot{\theta}_2 = -c_2\dot{\theta}_2 - k_2\theta_2 - cl\sin\left(\tilde{\theta}_2 - \theta_3\right)\dot{u} - kl\sin\left(\tilde{\theta}_2 - \theta_3\right)u + \tau_2\,, \tag{1}$$

$$I_4\ddot{\theta}_4 = -c_4\dot{\theta}_4 - k_4\theta_4 + cl\sin\left(\tilde{\theta}_4 - \theta_3\right)\dot{u} + kl\sin\left(\tilde{\theta}_4 - \theta_3\right)u + \tau_4\,, \tag{2}$$

where the subscripts 2 and 4 refer to the left and right joints, respectively. For the device here considered, the values of the parameters appearing in (1) and (2) are reported in Table 2.

Table 2. Numerical values of the parameters.

Parameter	Value	Unit
d	5.47×10^{-4}	m
l	1.50×10^{-3}	m
I_2, I_4	1.25×10^{-14}	$kg\,m^2$
k_2, k_4	0.30×10^{-6}	$N\,m^2$
c_2, c_4	1.24×10^{-12}	$N\,m^2\,s^{-1}$

The system has two degrees of freedom and it is fully described by Euqations (1) and (2). The variables θ_3 and u, and their time derivatives, can be computed as functions of the state variables θ_2, θ_4, $\dot{\theta}_2$ and $\dot{\theta}_4$ [34]:

$$\theta_3 = \arctan\frac{-l\sin\tilde{\theta}_2 + l\sin\tilde{\theta}_4}{d - l\cos\tilde{\theta}_2 + l\cos\tilde{\theta}_4}\,, \tag{3}$$

$$u = \sqrt{\left(d - l\cos\tilde{\theta}_2 + l\cos\tilde{\theta}_4\right)^2 + \left(-l\sin\tilde{\theta}_2 + l\sin\tilde{\theta}_4\right)^2} - \hat{u}\,, \tag{4}$$

$$\dot{u} = \dot{\theta}_2 l\sin\left(\tilde{\theta}_2 - \theta_3\right) - \dot{\theta}_4 l\sin\left(\tilde{\theta}_4 - \theta_3\right)\,. \tag{5}$$

4. Mechanical Characteristics Estimation

As briefly described in the Introduction, an approach to determine both the elastic and viscous characteristics of a biosample by using a microgripper was proposed in Reference [34], where particular waveforms were required for the input torques. However, this method is not suitable to perform simultaneous manipulation and measurement tasks.

To overcome this limitation, a measurement scheme adaptable to various operative conditions should be considered. Therefore, an on-line dynamical estimator could be implemented as a more efficient strategy for the mechanical characterization of the sample. Recursive least square methods can be successfully applied when the unknown parameters are the linear coefficients of the system dynamic equations. In such cases, quite usual in literature [43–46], the model can be rearranged as a linear time-varying system with respect to the parameters to be estimated, whereas the other terms, supposed measurable, are functions either of the state or of the output variables.

The model dynamics has to be rearranged as

$$y_i(t) = M_i(t)\omega_i(t)\,, \quad i \in [1,\ldots,m], \tag{6}$$

where m is the number of degrees of freedom of the system ($m = 2$ in the considered case), $\omega_i(t)$ is the vector of unknown parameters, and $y_i(t)$, $M_i(t)$ are known quantities depending on the system dynamics. It is worth noting that all the terms in Equation (6) are time dependent.

Equations (1) and (2) can be rewritten, according to the notation in Equation (6), in linear form with respect to the parameters as

$$y_i(t) = I_i\ddot{\theta}_i + c_i\dot{\theta}_i + k_i\theta_i - \tau_i,$$ (7)

$$M_i(t) = \mp l\sin(\theta_i + \hat{\theta}_i - \theta_3)\begin{pmatrix} \dot{u} & u \end{pmatrix},$$ (8)

$$\omega_i(t) = \begin{pmatrix} c(t) \\ k(t) \end{pmatrix},$$ (9)

where the sign in Equation (8) is minus or plus if $i = 2$ or if $i = 4$, respectively.

Referring to Equations (7)–(9), the general expressions to be defined for a generic recursive last squares (RLS) filtering algorithm are [47]:

$$\begin{aligned}
\hat{\omega}_i(t) &= \hat{\omega}_i(t-1) + K_i(t)\epsilon_i(t), \\
\epsilon_i(t) &= y_i(t) - \hat{y}_i(t), \\
\hat{y}_i(t) &= \phi_i^T(t)\hat{\omega}_i(t-1), \\
K_i(t) &= Q_i(t)\phi_i(t),
\end{aligned}$$ (10)

where $\hat{\omega}_i(t)$ and $\hat{y}_i(t)$ are the current estimation values of $\omega_i(t)$ and $y_i(t)$, $\epsilon_i(t)$ is the current prediction error, $K_i(t)$ is the gain determining how much the prediction error affects the update in the parameters estimation, and $\phi_i(t)$ represents the gradient of the predicted model output with respect to $\omega_i(t)$.

Two recursive last squares (RLS) filtering algorithms are applied considering different ranges of values for the parameters to be estimated. The algorithms are then compared to each other to understand how much the different viscous and elastic characteristics of the dynamical system affects the convergence of the estimation algorithms. With respect to the formulation in Equation (10), the two approaches differ for the choice of $Q_i(t)$. Moreover, it is worth noting that the symmetric structure of the dynamics represented by Equations (1) and (2) implies that, under a full state measurement, the results obtained choosing $i = 2$ are equal to the results corresponding to the case $i = 4$. Therefore, firstly the case $i = 2$ and consecutively the case $i = 4$ are addressed. Once equivalence and effectiveness of both cases has been proved, the choice in real applications should be driven by the simplicity of the measurement. For example, the operative technique of torque action on the first jaw only implies that for $i = 4$ no torque measurement is required, strongly simplifying the implementation.

4.1. Forgetting Factor Based RLS

The first method is based on a forgetting factor based RLS algorithm. In Equation (10), the following choice is performed

$$Q_i(t) = \frac{P_i(t-1)}{\lambda + \phi_i^T(t)P_i(t-1)\phi_i(t)},$$ (11)

where

$$P_i(t) = \frac{1}{\lambda}\left(P_i(t-1) - R_i(t)\right)$$ (12)

and

$$R_i(t) = \frac{P_i(t-1)\phi_i(t)\phi_i^T(t)P_i(t-1)}{\lambda + \phi_i^T(t)P_i(t-1)\phi_i(t)}.$$ (13)

It's assumed that the residual $\epsilon_i(t)$ (the difference between the estimated and the measured value of $y_i(t)$) is affected by a white noise with covariance equal to 1. According to previous equations, $\hat{\omega}_i(t)$ is computed in order to minimize the sum of residuals squares

$$\hat{\omega}_i(t) = \arg\min_{\theta} \sum_{k=1}^{t} \lambda^{t-k} \epsilon_2^i(k). \tag{14}$$

In Equations (11)–(14), $\lambda \in R$ is the forgetting factor introduced in order to consider differently the time sequence of the errors $\epsilon_i(t)$, according to an exponentially decreasing weight if $\lambda \in (0, 1)$. This choice is effective in case of time varying parameters while, when dealing with constant parameters, the choice $\lambda = 1$ is usually adopted.

The algorithm in Equation (10), with positions (11)–(13), has been applied to the case $i = 2$. In all the simulations, the initial values of the parameters have been chosen far from the real values. The initial covariance, proportional to P_2, has been fixed taking into account that the covariance matrix has to be chosen according to a priori knowledge of the parameters at $t = 0$: very high values of the covariance matrix elements correspond to completely unknown parameters.

Remark 1. *Note that the forgetting factor method is a particular simplified case of the Kalman filter.*

4.2. Normalised Gradient Based RLS

The second approach consists of a normalized gradient algorithm. This technique is based on the choice of $Q_i(t)$ with a simpler form with respect to (11):

$$Q_i(t) = \frac{\gamma}{|\phi_i(t)|^2 + \beta}, \tag{15}$$

where γ is the adaptation gain scaled by the gradient $\phi_i(t)$. The bias term β is added to the square norm of the gradient vector in the denominator of the expression, in order to prevent critical situations in the case $\phi_i(t)$ is close to zero. The algorithm (10) with $Q_i(t)$ defined by (15) requires only the initialization for the values of the parameters to be estimated.

Since the presence of noise is not explicitly considered in this formulation, the drawback is a smaller rate of convergence and a larger sensitivity to the presence of noises.

This technique has been simulated making reference to the dynamics represented by Equation (2), that is with $i = 4$.

5. Simulations

Numerical simulations, using Matlab® (MathWorks, Inc., Natick, MA, USA) and Simulink® (MathWorks, Inc., Natick, MA, USA) tools, were performed in order to show effectiveness, benefits and differences of the proposed estimation methods. Three different numerical cases were analysed, considering predominant elastic or dissipative behaviours.

The first case corresponds to a realistic condition with elastic and damping coefficient much greater than the ones of the mechanical structure, with $c = 8.4 \times 10^{-6}$ Nms/rad and $k = 2.5 \times 10^{-3}$ Nm/rad, where the elastic coefficient greater than the damping one.

In the second case, a damping coefficient greater than the elastic one was considered in order to check, by comparison, the dependency of the algorithm convergence from the two different mechanical characteristics. The order of magnitude for the two coefficients have been exchanged, setting $c = 8.4 \times 10^{-3}$ Nms/rad and $k = 2.5 \times 10^{-6}$ Nm/rad.

In the last case, a very poorly damped sample has been chosen with $c = 8.4 \times 10^{-11}$ Nms/rad, whereas $k = 2.5 \times 10^{-5}$ Nm/rad.

The initial parameters values, for all the simulations and for both the estimation algorithms have been chosen as $c(0) = 10^{-9}$ Nms/rad and $k(0) = 10^{-7}$ Nm/rad.

For the forgetting factor RLS estimator introduced in Subsection 4.1, the 2×2 square covariance matrix is diagonal, with both the diagonal elements equal to 10^{20}, while the forgetting factor λ is fixed to $\lambda = 0.99$.

For the normalized gradient estimator, presented in Subsection 4.2, the adaptation gain γ has been set as $\gamma = 0.9$ and the normalization bias has been chosen as $\beta = 2.2 \times 10^{-16}$.

Simulation results obtained for the first case ($c = 8.4 \times 10^{-6}$ Nms/rad and $k = 2.5 \times 10^{-3}$ Nm/rad) are depicted in Figure 4a for the elastic coefficient k and in Figure 4b for the damping coefficient c. The solid line shows the estimation evolution with the first algorithm, whereas the dashed one reports the results of the second algorithm. The dotted line corresponds to the true values of the parameters, plotted as a reference. As expected, both the algorithms converge in a very short time, but the first is faster than the second.

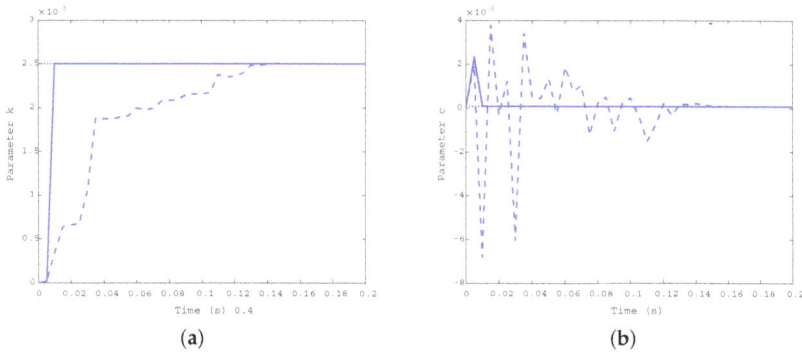

Figure 4. Time evolution of the estimated parameters k (**a**) and c (**b**) for the first case: $c = 8.4 \times 10^{-6}$ Nms/rad, $k = 2.5 \times 10^{-3}$ Nm/rad.

For the second case ($c = 8.4 \times 10^{-3}$ Nms/rad and $k = 2.5 \times 10^{-6}$ Nm/rad), the simulation results are depicted in Figure 5a for the damping coefficient c, and in Figure 5b for the elastic one k. As in the previous case, the solid line refers to the estimation evolution with the first algorithm, the dashed one refers to the second algorithm, and the dotted one is the true reference value. The difference in the convergence rate for the two approaches is confirmed.

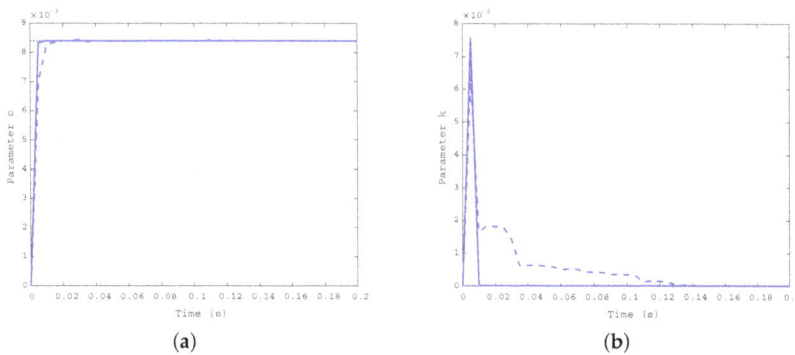

Figure 5. Time evolution of the estimated parameters c (**a**) and k (**b**) for the second case: $c = 8.4 \times 10^{-3}$ Nms/rad, $k = 2.5 \times 10^{-6}$ Nm/rad.

The results obtained by simulation of the third case ($c = 8.4 \times 10^{-11}$ Nms/rad and $k = 2.5 \times 10^{-5}$ Nm/rad) are reported in Figure 6a for the elastic coefficient k, and in Figure 6b for the damping one c. The difference in the convergence rate for the two approaches is confirmed, and uniformity in the estimation of the damping coefficient c with the two algorithms is the same as in the first case.

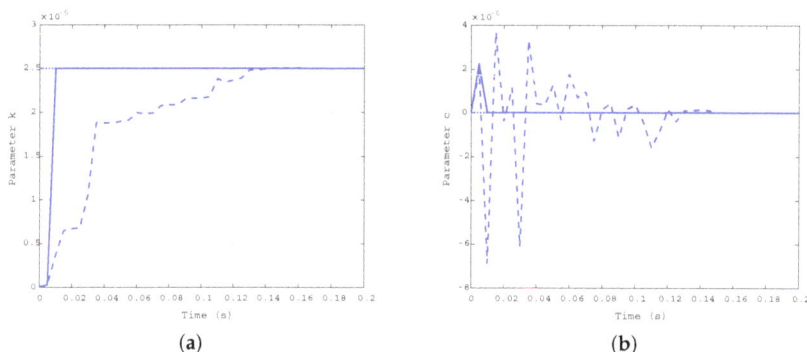

(a)   (b)

Figure 6. Time evolution of the estimated parameters k (**a**) and c (**b**) for the third case: $c = 8.4 \times 10^{-11}$ Nms/rad, $k = 2.5 \times 10^{-5}$ Nm/rad.

Numerical results support the effectiveness of the proposed method, tested considering different orders of magnitude for the parameters to be estimated. Clearly, in the present formulation, the state is assumed to be measurable, with possible additive Gaussian noise to model uncertainties and measurement noise.

6. Conclusions

In this paper, the possibility of using a microgripper for the identification of the mechanical properties of biomaterials was investigated. To describe the gripper–sample system dynamics, the pseudo-rigid body model and the Kelvin–Voigt constitutive law of viscoelasticity were considered for the microgripper and for the sample, respectively. The normalised gradient-based RLS and the forgetting factor based RLS algorithms were implemented to solve the parameters estimation problem. Three cases were analysed, assuming different ranges of values for the coefficients to be estimated. The simulation results confirmed the feasibility of the method: both the algorithms converged in a very short time, but the normalised gradient-based RLS algorithm resulted to be faster than the other one. Therefore, the proposed approach could enable the ability to simultaneously perform the manipulation of the biosample and the identification of its mechanical characteristics, i.e., stiffness and damping coefficients. However, the application of the proposed method to a variety of biological samples is limited by the assumption of the Kelvin–Voigt constitutive law of viscoelasticity. From the methodological point of view, further investigations will focus on the implementation of more complex constitutive models and on the design of high-performance algorithms, always attaining robustness with respect to noise and parameter uncertainties.

Author Contributions: P.D.G. and M.L.A. implemented the estimation algorithms and performed the simulations. O.G. and M.V. developed the model adopted for the kinematic and dynamic analysis of the system. All authors contributed in writing the manuscript.

Funding: This research received no external funding.

Acknowledgments: The Authors acknowledge the Micro Nano Facility, Center for Materials and Microsystems of Fondazione Bruno Kesler (FBK-CMM-MNF) for the fabrication of the device and the technical support (Figure 1).

Conflicts of Interest: The authors declare no conflict of interest.

References

1. Backman, D.E.; LeSavage, B.L.; Shah, S.B.; Wong, J.Y. A Robust Method to Generate Mechanically Anisotropic Vascular Smooth Muscle Cell Sheets for Vascular Tissue Engineering. *Macromol. Biosci.* **2017**, *17*. [CrossRef] [PubMed]
2. Mijailovic, A.S.; Qing, B.; Fortunato, D.; Van Vliet, K.J. Characterizing viscoelastic mechanical properties of highly compliant polymers and biological tissues using impact indentation. *Acta Biomater.* **2018**, *71*, 388–397. [CrossRef] [PubMed]
3. Qasaimeh, M.; Sokhanvar, S.; Dargahi, J.; Kahrizi, M. A micro-tactile sensor for in situ tissue characterization in minimally invasive surgery. *Biomed. Microdevices* **2008**, *10*, 823–837. [CrossRef] [PubMed]
4. Zhang, W.; Liu, L.F.; Xiong, Y.J.; Liu, Y.F.; Yu, S.B.; Wu, C.W.; Guo, W. Effect of in vitro storage duration on measured mechanical properties of brain tissue. *Sci. Rep.* **2018**, *8*, 1247. [CrossRef] [PubMed]
5. Laksari, K.; Assari, S.; Seibold, B.; Sadeghipour, K.; Darvish, K. Computational simulation of the mechanical response of brain tissue under blast loading. *Biomech. Model. Mechanobiol.* **2015**, *14*, 459–472. [CrossRef] [PubMed]
6. Vivanco, J.; Aiyangar, A.; Araneda, A.; Ploeg, H.L. Mechanical characterization of injection-molded macro porous bioceramic bone scaffolds. *J. Mech. Behav. Biomed. Mater.* **2012**, *9*, 137–152. [CrossRef] [PubMed]
7. Prasadh, S.; Wong, R.C.W. Unraveling the mechanical strength of biomaterials used as a bone scaffold in oral and maxillofacial defects. *Oral Sci. Int.* **2018**, *15*, 48–55. [CrossRef]
8. Edsberg, L.E.; Cutway, R.; Anain, S.; Natiella, J.R. Microstructural and mechanical characterization of human tissue at and adjacent to pressure ulcers. *J. Rehabil. Res. Dev.* **2000**, *37*, 463–471. [PubMed]
9. Hsu, C.K.; Lin, H.H.; Hans, I.; Harn, C.; Hughes, M.; Tang, M.J.; Yang, C.C. Mechanical forces in skin disorders. *J. Dermatol. Sci.* **2018**, *90*, 232–240. [CrossRef] [PubMed]
10. Whitford, C.; Movchan, N.V.; Studer, H.; Elsheikh, A. A viscoelastic anisotropic hyperelastic constitutive model of the human cornea. *Biomech. Model. Mechanobiol.* **2018**, *17*, 19–29. [CrossRef] [PubMed]
11. Erath, B.D.; Zañartu, M.; Peterson, S.D. Modeling viscous dissipation during vocal fold contact: the influence of tissue viscosity and thickness with implications for hydration. *Biomech. Model. Mechanobiol.* **2017**, *16*, 947–960. [CrossRef] [PubMed]
12. Garcés-Schröder, M.; Metz, D.; Hecht, L.; Iyer, R.; Leester-Schädel, M.; Böl, M.; Dietzel, A. Characterization of skeletal muscle passive mechanical properties by novel micro-force sensor and tissue micro-dissection by femtosecond laser ablation. *Microelectron. Eng.* **2018**, *192*, 70–76. [CrossRef]
13. Huveneers, S.; Daemen, M.J.; Hordijk, P.L. Between Rho (k) and a hard place: the relation between vessel wall stiffness, endothelial contractility, and cardiovascular disease. *Circ. Res.* **2015**, *116*, 895–908. [CrossRef] [PubMed]
14. Pogoda, K.; Chin, L.; Georges, P.C.; Byfield, F.J.; Bucki, R.; Kim, R.; Weaver, M.; Wells, R.G.; Marcinkiewicz, C.; Janmey, P.A. Compression stiffening of brain and its effect on mechanosensing by glioma cells. *New J. Phys.* **2014**, *16*, 075002. [CrossRef] [PubMed]
15. Hu, S.; Yang, C.; Hu, D.; Lam, R.H. Microfluidic biosensing of viscoelastic properties of normal and cancerous human breast cells. In Proceedings of the 2017 IEEE 12th International Conference on Nano/Micro Engineered and Molecular Systems (NEMS), Los Angeles, CA, USA, 9–12 April 2017; pp. 90–95.
16. Zouaoui, J.; Trunfio-Sfarghiu, A.; Brizuela, L.; Piednoir, A.; Maniti, O.; Munteanu, B.; Mebarek, S.; Girard-Egrot, A.; Landoulsi, A.; Granjon, T. Multi-scale mechanical characterization of prostate cancer cell lines: Relevant biological markers to evaluate the cell metastatic potential. *Biochim. Biophys. Acta Gen. Subj.* **2017**, *1861*, 3109–3119. [CrossRef] [PubMed]
17. Rubiano, A.; Delitto, D.; Han, S.; Gerber, M.; Galitz, C.; Trevino, J.; Thomas, R.M.; Hughes, S.J.; Simmons, C.S. Viscoelastic properties of human pancreatic tumors and in vitro constructs to mimic mechanical properties. *Acta Biomater.* **2018**, *67*, 331–340. [CrossRef] [PubMed]
18. Kauer, M.; Vuskovic, V.; Dual, J.; Székely, G.; Bajka, M. Inverse finite element characterization of soft tissues. *Med. Image Anal.* **2002**, *6*, 275–287. [CrossRef]
19. Dargahi, J.; Najarian, S. Advances in tactile sensors design/manufacturing and its impact on robotics applications—A review. *Ind. Robot* **2005**, *32*, 268–281. [CrossRef]

20. Dargahi, J.; Najarian, S.; Mirjalili, V.; Liu, B. Modelling and testing of a sensor capable of determining the stiffness of biological tissues. *Can. J. Electr. Comput. Eng.* **2007**, *32*, 45–51. [CrossRef]

21. Addae-Mensah, K.A.; Wikswo, J.P. Measurement techniques for cellular biomechanics in vitro. *Exp. Biol. Med.* **2008**, *233*, 792–809. [CrossRef] [PubMed]

22. Rodriguez, M.L.; McGarry, P.J.; Sniadecki, N.J. Review on cell mechanics: Experimental and modeling approaches. *Appl. Mech. Rev.* **2013**, *65*, 060801. [CrossRef]

23. Chen, K.; Yao, A.; Zheng, E.E.; Lin, J.; Zheng, Y. Shear wave dispersion ultrasound vibrometry based on a different mechanical model for soft tissue characterization. *J. Ultrasound Med.* **2012**, *31*, 2001–2011. [CrossRef] [PubMed]

24. Lim, C.; Zhou, E.; Quek, S. Mechanical models for living cells—A review. *J. Biomech.* **2006**, *39*, 195–216. [CrossRef] [PubMed]

25. Johnson, M.L.; Faunt, L.M. Parameter estimation by least-squares methods. In *Methods in Enzymology*; Elsevier: Amsterdam, The Netherlands 1992; Volume 210, pp. 1–37.

26. Xi, J.; Lamata, P.; Lee, J.; Moireau, P.; Chapelle, D.; Smith, N. Myocardial transversely isotropic material parameter estimation from in-silico measurements based on a reduced-order unscented Kalman filter. *J. Mech. Behav. Biomed. Mater.* **2011**, *4*, 1090–1102. [CrossRef] [PubMed]

27. Boonvisut, P.; Çavuşoğlu, M.C. Estimation of soft tissue mechanical parameters from robotic manipulation data. *IEEE/ASME Trans. Mechatron.* **2013**, *18*, 1602–1611. [CrossRef] [PubMed]

28. Yang, S.; Xu, Q.; Nan, Z. Design and development of a dual-axis force sensing MEMS microgripper. *J. Mech. Robot.* **2017**, *9*, 061011. [CrossRef]

29. Verotti, M.; Dochshanov, A.; Belfiore, N.P. A comprehensive survey on microgrippers design: Mechanical structure. *J. Mech. Des.* **2017**, *139*, 060801. [CrossRef]

30. Dochshanov, A.; Verotti, M.; Belfiore, N.P. A comprehensive survey on microgrippers design: Operational strategy. *J. Mech. Des.* **2017**, *139*, 070801. [CrossRef]

31. Cauchi, M.; Grech, I.; Mallia, B.; Mollicone, P.; Sammut, N. Analytical, Numerical and Experimental Study of a Horizontal Electrothermal MEMS Microgripper for the Deformability Characterisation of Human Red Blood Cells. *Micromachines* **2018**, *9*, 108. [CrossRef]

32. Velosa-Moncada, L.A.; Aguilera-Cortés, L.A.; González-Palacios, M.A.; Raskin, J.P.; Herrera-May, A.L. Design of a Novel MEMS Microgripper with Rotatory Electrostatic Comb-Drive Actuators for Biomedical Applications. *Sensors* **2018**, *18*, 1664. [CrossRef] [PubMed]

33. Potrich, C.; Lunelli, L.; Bagolini, A.; Bellutti, P.; Pederzolli, C.; Verotti, M.; Belfiore, N.P. Innovative Silicon Microgrippers for Biomedical Applications: Design, Mechanical Simulation and Evaluation of Protein Fouling. *Actuators* **2018**, *7*, 12. [CrossRef]

34. Di Giamberardino, P.; Bagolini, A.; Bellutti, P.; Rudas, I.J.; Verotti, M.; Botta, F.; Belfiore, N.P. New MEMS Tweezers for the Viscoelastic Characterization of Soft Materials at the Microscale. *Micromachines* **2018**, *9*, 15. [CrossRef]

35. Verotti, M.; Crescenzi, R.; Balucani, M.; Belfiore, N.P. MEMS-based conjugate surfaces flexure hinge. *J. Mech. Des.* **2015**, *137*, 012301. [CrossRef]

36. Verotti, M.; Dochshanov, A.; Belfiore, N.P. Compliance synthesis of CSFH MEMS-based microgrippers. *J. Mech. Des.* **2017**, *139*, 022301. [CrossRef]

37. Cecchi, R.; Verotti, M.; Capata, R.; Dochshanov, A.; Broggiato, G.B.; Crescenzi, R.; Balucani, M.; Natali, S.; Razzano, G.; Lucchese, F.; et al. Development of micro-grippers for tissue and cell manipulation with direct morphological comparison. *Micromachines* **2015**, *6*, 1710–1728. [CrossRef]

38. Bagolini, A.; Ronchin, S.; Bellutti, P.; Chiste, M.; Verotti, M.; Belfiore, N. Fabrication of novel MEMS microgrippers by deep reactive ion etching with metal hard mask. *IEEE J. Microelectromech. Syst.* **2017**, *26*, 926–934. [CrossRef]

39. Bagolini, A.; Bellutti, P.; Di Giamberardino, P.; Rudas, I.J.; D'Andrea, V.; Verotti, M.; Dochshanov, A.; Belfiore, N. Stiffness characterization of biological tissues by means of MEMS-technology based micro grippers under position control. *Mech. Mach. Sci.* **2018**, *49*, 939–947.

40. Verotti, M. Analysis of the center of rotation in primitive flexures: Uniform cantilever beams with constant curvature. *Mech. Mach. Theory* **2016**, *97*, 29–50. [CrossRef]

41. Verotti, M. Effect of initial curvature in uniform flexures on position accuracy. *Mech. Mach. Theory* **2018**, *119*, 106–118. [CrossRef]

42. Sanò, P.; Verotti, M.; Bosetti, P.; Belfiore, N.P. Kinematic Synthesis of a D-Drive MEMS Device with Rigid-Body Replacement Method. *J. Mech. Des.* **2018**, *140*, 075001. [CrossRef]

43. Flacco, F.; De Luca, A.; Sardellitti, I.; Tsagarakis, N.G. Robust estimation of variable stiffness in flexible joints. In Proceedings of the 2011 IEEE/RSJ International Conference on Intelligent Robots and Systems, San Francisco, CA, USA, 25–30 September 2011.

44. Lundquist, C.; Schön, T.B. Recursive Identification of Cornering Stiffness Parameters for an Enhanced Single Track Model. *IFAC Proc. Vol.* **2009**, *42*, 1726–1731. [CrossRef]

45. Vahidi, A.; Stefanopoulou, A.; Peng, H. Recursive least squares with forgetting for online estimation of vehicle mass and road grade: Theory and experiments. *Veh. Syst. Dyn.* **2005**, *43*, 31–55. [CrossRef]

46. Lee, S.D.; Jung, S. A recursive least square approach to a disturbance observer design for balancing control of a single-wheel robot system. In Proceedings of the 2016 IEEE International Conference on Information and Automation (ICIA), Ningbo, China, 31 July–4 August 2016; pp. 1878–1881.

47. Ljung, L. *System Identification: Theory for the User*, 2nd ed.; Prentice Hall: Upper Saddle River, NJ, USA, 1999.

actuators

Article

Preparing and Mounting Polymer Nanofibers onto Microscale Test Platforms

Ramesh Shrestha, Sheng Shen and Maarten P. de Boer *

Mechanical Engineering Department, Carnegie Mellon University, Pittsburgh, PA 15217, USA;
rshresth@andrew.cmu.edu (R.S.); sshen1@cmu.edu (S.S.)
* Correspondence: mpdebo@andrew.cmu.edu

Received: 7 September 2018; Accepted: 10 October 2018; Published: 15 October 2018

Abstract: Because they can achieve a high degree of molecular chain alignment in comparison with their bulk counterparts, the mechanical and thermal properties of polymer nanofibers are of great interest. However, due to their nanometer-scale size, it is difficult to manipulate, grip, and test these fibers. Here, we demonstrate simple repeatable methods to transfer as-drawn fibers to micrometer-scale test platforms where their properties can be directly measured. Issues encountered and methods to minimize measurement artifacts are also discussed.

Keywords: polymer nanofiber; manipulation; precision mounting; UHMWPE

1. Introduction

Polymer nanofibers have mechanical and thermal properties significantly higher than their bulk counterparts [1–6]. They have a wide range of potential applications in areas such as tissue engineering, sensing, reinforcing composites, ballistic armor, textiles, and in improving thermal management [7–11]. Detailed study of their material properties is necessary to assess their true application possibilities.

However, due to their nanometer-scale size, it is difficult to manipulate and mount individual polymer nanofibers onto a test platform to make measurements. The macroscale approach of manipulating objects with pick-and-place tools has been scaled to smaller sizes using micro- and nanoscale grippers [12–14]. A tweezer-like gripping mechanism in these grippers can be obtained using various actuation mechanisms such as electrostatic, thermal, shape memory, fluidic, magnetic, and piezoelectric [14]. But it is difficult to ensure reliable placement of a long nanofiber onto a horizontal platform using a nanogripper. Optical tweezing is capable of sorting and manipulating an individual DNA molecule. This technique requires a liquid medium, which is not compatible with mechanical measurement platforms operating in laboratory air [15]. Another common approach is to use an atomic force microscope (AFM) cantilever tip or a tungsten tip mounted on a precision XYZ manipulator. This has been successfully used for nanoscale fabrication of polymer nanofibers [16] and precision mounting of nanofibers [1,17]. This approach is low yield and time-consuming. In addition, the process is generally performed in a scanning electron microscope (SEM) [17] because nanofibers are difficult to resolve optically. However, polymer nanofibers are damaged by electron-beam irradiation [18]. Hence, reliable optical methods are strongly needed for manipulation and testing.

In view of the urgent need for such micromanipulation, this article describes and demonstrates a simple specimen collector technique that enables nanofiber placement with ±1 μm precision onto micromachined test platforms using optical methods. Electrospinning is one common method to make such long fibers [9,19,20]. Here, as an alternative example, we describe a method by which ultra-high molecular weight polyethylene (UHMWPE) nanofibers can be made. Then we show how they can be manipulated onto micromachined test platforms without damage. Next, we demonstrate a nanomechanical test calibration procedure, discuss potential testing artifacts in nanomechanical and nano thermal measurements, and propose methods to solve them. Detailed characterization, thermal and nanomechanical measurements on highly crystalline nanofibers prepared by this method have been reported in reference [3].

2. Experimental

2.1. Ultra-High Molecular Weight Polyethylene (UHMWPE) Nanofiber Specimen Preparation

Ultra-high molecular weight polyethylene (UHMWPE), a linear chain polymer, is used in this work. However, the processes described here can be generalized to other drawable polymers. To fabricate the nanofibers, we first produced a gel from 0.8 wt % UHMWPE powder (average molecular weight ~3×10^6–6×10^6 g/mol purchased from Sigma Aldrich, St. Louis, MO, USA) and mixed it with decalin solvent [21]. The procedure was carried out inside an argon-filled glove box to avoid oxidation and subsequent molecular degradation. After heating on a hot plate to 145 °C, the mixture became transparent and viscous as the UHMWPE powder dissolved in the solvent while a glass rod constantly stirred the solution. The solution was then quenched in a room temperature water bath, and the gel formed.

Next, we devised a two-stage tip drawing method. First a μm-scale UHMWPE fiber (MF) was drawn, as schematically represented in Figure 1. There, a 5 mm × 5 mm silicon chip with a thin film heater attached on the backside heated the gel to 120–130 °C. The temperature was measured using a thermocouple attached to the thin film heater. A hot plate placed 1 cm below the silicon chip heated the overhead air to 90 °C to prepare for hot-stretching. The translucent gel transformed to a transparent solution as it reached 130 °C. As the solution became clear, a sharp (~10 μm) glass tip was inserted into the solution using an XYZ-axis micromanipulator. It drew a MF of short length (several hundreds of μm). The drawing induces stress-induced crystallization with folded chain lamellae (kebab) on oriented chains (sheesh). Upon drawing, the lamellar structure unfolds into microfibrils that are pulled taut between entanglements [4]. During crystallization, the decalin evaporates, aided by the convective current from the hot plate. Decalin syneresis further facilitates crystallization. The fiber was then further drawn to a length of 1 cm, after which it was quenched to room temperature to minimize relaxation of the extended chain.

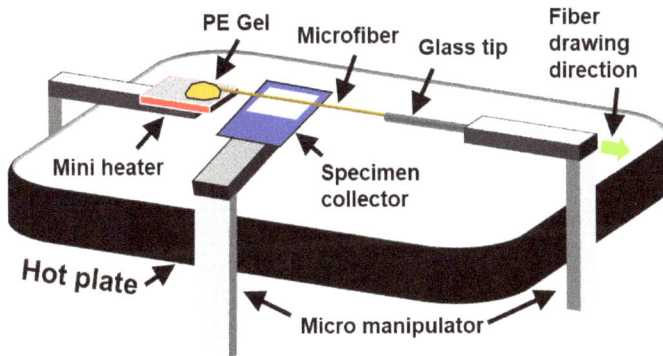

Figure 1. Schematic of tip drawing setup.

To manipulate and further process the microfiber, a bulk-micromachined specimen collector was fabricated by bulk micromachining processes, as detailed in Appendix A. It is a 1 cm × 3 cm silicon chip of 525 μm thickness and a square hole of 7.5 mm × 7.5 mm. Again using a standard XYZ-axis micromanipulator, the specimen collector was raised from under the drawn MF, as depicted in Figure 1, and the MF was gripped onto the specimen collector using epoxy. Because it was stretched when the specimen collector was raised and glued, the MF remained taut with a moderate tensile stress. The MF was then cut outside the collector edges.

With the specimen collector acting as a carrier, the MF could now be conveniently transferred from the drawing station and positioned under an optical microscope objective. There, a locally etched and bent tungsten wire, used as a microheater, was placed in the same horizontal plane 20 μm from the MF. Details of the microheater construction are given in Appendix B.

A voltage was applied to the microheater, the tungsten wire temperature increased due to Joule heating, and the collected MF was heated during a ~1 s pulse by convection to make a UHMWPE *nanofiber* (NF), as schematically shown in Figure 2. Most advanced thermometers have either temporal or spatial resolution limitations that limit direct temperature measurement during this process. We estimated the maximum temperature attained by the fiber during local heating using the ANSYS/FLUENT computational fluid dynamics finite element analysis software package [22]. First, the microheater temperature profile was calculated. Second, we assumed a MF with a typical 3 μm diameter and placed at a distance of 20 μm from the fiber. A potential bias 0.7 V was applied to the heater, as in the experiment. Both convective and radiative heat losses were considered. At the microscale, the boundary layer is smaller than at the macroscale. Thus, the heat transfer coefficient (h) greatly increases. Here, we used $h = 1000 \text{ W m}^{-2} \text{ K}^{-1}$ and emissivity = 0.1 [23]. Figure 3a shows the simulated temperature MF profile. The thermal conductivity of the MF was modeled using gel spun microfibers with $k = 20 \text{ W m}^{-1} \text{ K}^{-1}$ [24]. We used an emissivity value of the MF of 0.2 [25]. A constant thermal conductivity of air, $k = 0.02 \text{ W m}^{-1} \text{ K}^{-1}$, was used. The MF temperature reached 450 K (180 °C), and the temperature profile along the fiber is shown in Figure 3b.

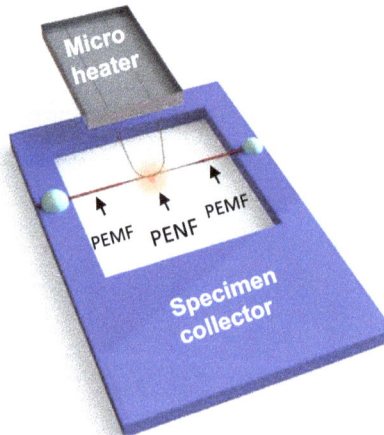

Figure 2. Schematic of the specimen collector and the fabrication of an ultra-high molecular weight polyethylene (UHMWPE) nanofiber (NF). Adapted from reference [3] supplementary information.

During this short temperature cycle, the MF temperature likely approached the melting temperature, and stretched rapidly due to the residual tension mentioned above. As the NF developed, its molecular alignment and crystallization was further increased. This highly aligned structure was then quenched as the NF rapidly cooled after the heater was turned off. The NF, now suspended between MF regions outside the heater area, was itself typically no longer optically observable, but the outer MF regions resided in the same location as before the second drawing stage. This indicated that a NF now had been created and still physically connected the MF regions, as represented in Figure 2.

The geometrical dimensions of NFs created by this fabrication sequence were characterized by scanning electron microscopy (SEM). We found that factors influencing the diameter included the initial MF diameter, the pre-existing tensile stress, and the heater current. The diameters ranged from

20–250 nm and a length of 100 µm. Detailed specimen geometry information is provided in reference [3]. Further characterization data on twinning as determined from transmission electron microscopy and on molecular alignment as inferred from Raman spectroscopy has also been reported [3].

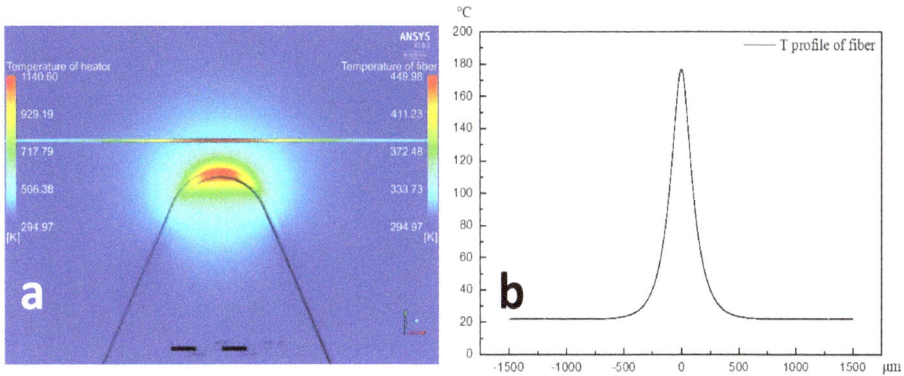

Figure 3. Temperature profile of the W wire heater and the polyethylene microfiber (MF) from an ANSYS/FLUENT simulation. (**a**) Temperature map of the heater and the MF. (**b**) Plot of the MF temperature profile. The maximum temperature reaches 450 K (180 °C). Adapted from reference [3] supplementary information.

2.2. Positioning the Nanofibers on the Test Platforms

2.2.1. Mechanical Test Platform

A micromachined stepper motor, detailed in [26], was selected for the mechanical measurements. The test platform was designed at Carnegie Mellon University and microfabricated by Sandia National Labs, Albuquerque, NM, USA using the SUMMiT V process [27]. The polycrystalline silicon (polysilicon) motor was "released" in the Carnegie Mellon University nanofabrication facility. The release process involves removing silicon oxide layers in hydrofluoric acid. Then a critical point drying process [28] rendered the structure freestanding. The full platform included a trailing load cell and grips, as seen on the left of Figure 4. The motor was actuated electrostatically, and each step was approximately 60 nm. Here the steps were taken at 1 Hz. It can generate an in-plane force of up to 1 mN, substantially more than the ~100 µN needed to stretch and break the NFs. Subpixel interpolation with approximately 5 nm resolution was used to measure displacements, as recorded through a 50× objective in an optical microscope. Because the load spring and the NF specimen are in series, the displacement in each depends on their relative stiffnesses. Here, the load spring stiffness was on the same order as the NFs. The tensile load was estimated by measuring the separation between the load cell combs, while the specimen extension was determined from the comb attached to the moving grip.

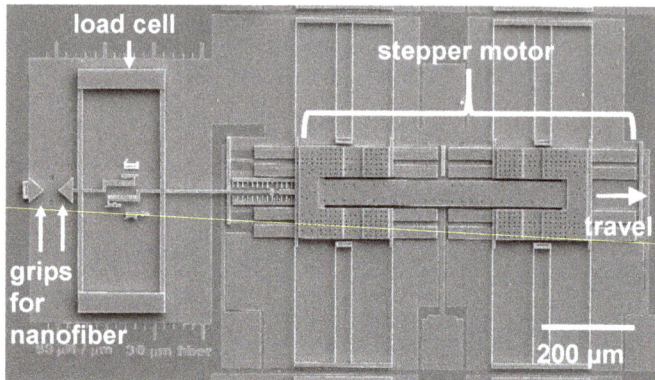

Figure 4. The micromachined test platform used for mechanical measurements.

The minuscule sample size, the limited resolution of the optical microscope and nonlinear forces such as van der Waals, triboelectric, and capillary forces due to moisture are factors that make it difficult to successfully place nanofibers with tip-based methods. Yet, it is critical to locate and position the nanofiber precisely onto the micromachined test platform without damaging it. Electron microscopes cannot be used because of the sensitivity of polymer samples to electron beam radiation [29,30], which scissions bonds and hence degrade the mechanical and thermal property enhancement [31,32]. The manipulation by optical methods is preferred. Regarding alignment, during tensile testing, an unwanted bending moment can cause the load cell to rotate and the fiber force will be underestimated.

The specimen collector with a suspended NF was again used as a specimen carrier, now to align the NF with the mechanical test apparatus. It was maneuvered above the grips, as shown in Figure 5. The suspended MF in the undrawn region was used as a reference to align the nanofiber. Two triangular pads, which must grip the specimen, are behind the load spring, as seen in Figure 4. One pad is fixed to the substrate while the other is connected to the load cell. The first step in mounting the nanofiber involves pre-gluing. A small volume of cyanoacrylate glue (~20 μm × 5 μm × 5 μm) was transferred to the load cell beam that is connected to the specimen displacement-indicating comb using a compliant tungsten tip. This was maneuvered into position with a precise motorized micromanipulator. This glue held the specimen in place while the fiber was aligned and cut. The specimen collector was held by a second micromanipulator and was aligned along triangular pads under a 50× optical objective. This was done quickly (<5 min), as the glue cured in 20 min. The specimen collector was then lowered until the fiber made contact with the glue. It was fixed at that position until the glue cured enough to hold the fiber while cutting. Figure 6 shows a pre-glued PE nanofiber aligned to the grips. In this case, as in other images of fibers presented in this article, the fiber diameter is ≈250 nm so that it is optically detectable. Fibers with diameters <100 nm were also successfully mounted. Though they could not be observed optically, their presence was confirmed during mechanical or thermal testing.

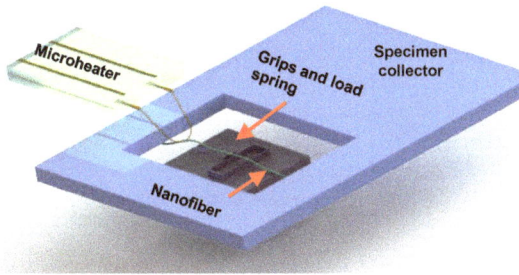

Figure 5. Schematic of alignment of the NF along the platform grips (figure not to scale). Adapted from reference [3] supplementary information.

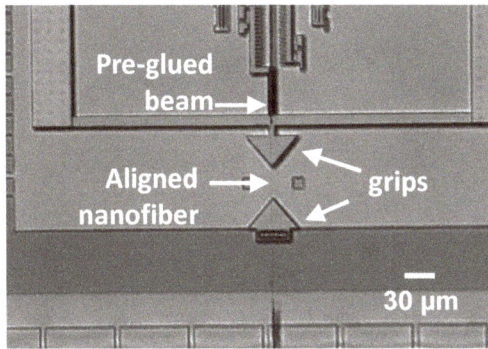

Figure 6. Optical image showing the pre-glued beam and an aligned nanofiber.

Next, the fibers were glued to the grips. Using a sharp compliant tungsten tip, glue was gently placed onto the aligned fiber over the pads. If the fiber came out of alignment, it could be nudged back into position using the same tungsten tip. As the glue became more viscous, manipulating it became more difficult and the success rate diminished. Since the working time of the glue was ~8 min, this process was completed within 6–8 min. Figure 7 shows a sample that has successfully been gripped at both ends.

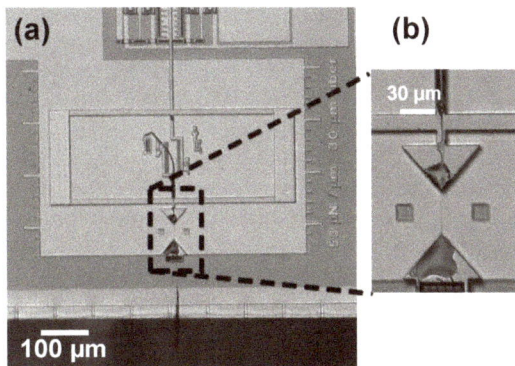

Figure 7. Optical image of UHWMPE nanofiber tor ready for mechanical testing. (**a**) The fiber is glued to both grips pads after the cutting process (**b**) Magnified image of the pads after gluing.

After 15 min, the glue had partially cured. The microheater then cut the fiber to a length of 200–300 µm by Joule heating, much longer than the 30 µm gauge length. A DC power of 8–15 mW cut the fibers. As the heater made contact with the fiber, it melted it locally, thereby cutting it. Due to the tensile stress remaining from the specimen collector, the fiber was pulled towards the pad by a few µm. Similarly, the other end of the fiber was cut, thereby keeping the NF within the grips, with some extra length extending beyond the glue.

After a fiber had been glued, it was cured in laboratory air for 48 h, and was then ready for testing. Figure 8 shows a SEM image of a gripped nanofiber in the mechanical platform.

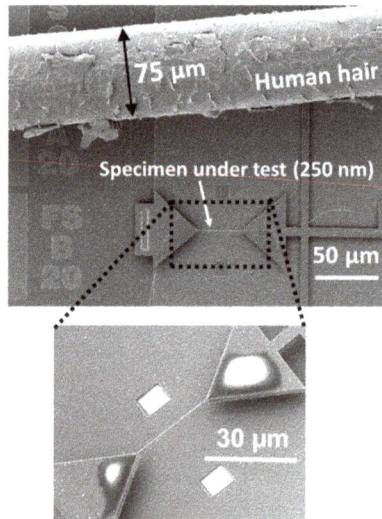

Figure 8. SEM image of a 250 nm diameter NF in comparison with human hair. Inset: detailed view of gripped nanofiber.

2.2.2. Thermal Test Platform

A different microdevice, consisting of suspended platinum resistance thermometers (PRTs) was used to conduct thermal characterization [33]. A total of seven SiN_x beams, with platinum deposited on each, supports each of two suspended SiN_x plates. Serpentine platinum coils are fabricated on top of the SiN_x plates and are connected to electrical pads in a supported region. A lock-in amplifier detects small changes in the resistance of these PRTs. Consequently, the device is sensitive to minute temperature changes. A NF connecting one platform to the adjacent one conducts heat from the heated side to the sensing side. Details of the measurement system are given in reference [33].

The alignment of the nanofiber to the plates was similar to that discussed above for the mechanical device. However, it was found that epoxy glues wetted the Pt wire surface, creating an insulating layer that significantly increased the thermal contact resistance, R_c. Without glue, the cutting process tended to pull the nanofiber off the suspended plates. Focused ion beam (FIB)-deposited Pt would likewise reduce the crystallinity and significantly increase the thermal resistance.

To solve this problem, we took advantage of the surface tension from evaporating isopropanol (IPA), which was able to affix the fiber onto the substrate. As the liquid evaporated, it pulled the fiber into contact with the plates and created sufficient adhesion to keep the nanofiber in place during the cutting process. This adhesion process was termed "capillary-assisted adhesion" and is discussed next.

Steps to achieve capillary-assisted adhesion are schematically shown in Figure 9. In this process, an IPA liquid drop was first placed on top of the thermal device (Figure 9b). Evaporation of IPA from the device was closely monitored under the optical microscope. After the liquid meniscus connecting the suspended plates broke, the suspending beams (not shown) cause the plates to separate to their nominal 5–10 µm gap. Before the liquid layer evaporated from the surface of the plates, a pre-aligned nanofiber was expeditiously lowered into focus and contact was made with the pads (Figure 9c). The liquid wet the fiber and the evaporating liquid surface pulled the fiber down due to surface tension. The fiber then adhered to the pads (Figure 9d). This surface tension, accompanied by the van der Waals force, increased the contact width of the fiber on the pads. Also, the fiber conformed to the undulating surface, increasing the contact length significantly. Increased contact width and contact length due to capillary-assisted adhesion reduced the thermal contact resistance. Figure 10 shows an optical image of a mounted nanofiber on the device. NFs that are of a significantly smaller diameter and not resolvable under the optical microscope could be manipulated and mounted onto the test platform.

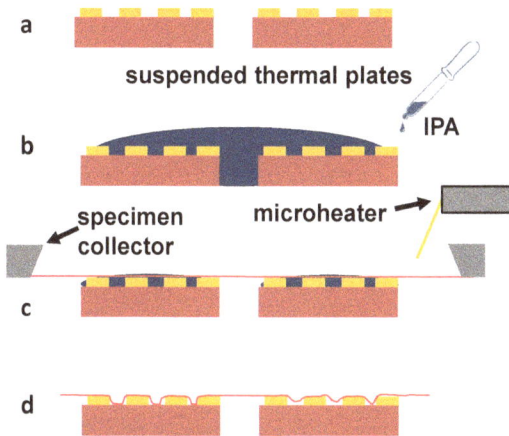

Figure 9. Schematic diagrams representing nanofiber placement onto a platinum resistance thermometer plates. (**a**) suspended plates, (**b**) a small drop of isopropanol (IPA) is placed on the platforms (**c**) after the IPA evaporates, the microheater cuts the fibers, (**d**) The evaporating IPA pulls the fiber into contact with the plate.

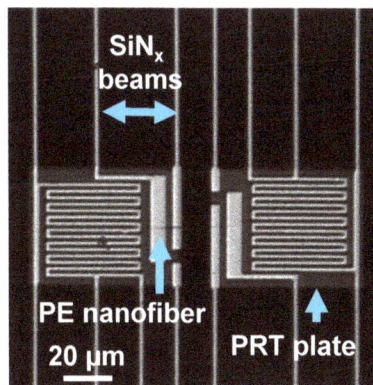

Figure 10. Optical image of a UHWMPE nanofiber sample mounted on the thermal device.

3. Results and Discussion

3.1. Calibration of Mechanical Test Platform Using Silica Fibers

Silica nanofibers were used to validate the nanomechanical measurements reported in [3]. They were prepared using a flame brushing technique [34], where a silica microwire was heated in the center using a butane torch and pulled at other two ends. It was stretched until broken; nanofibers were then obtained. The nanofiber was placed on a PDMS substrate, and in this case, conventional manipulation methods were used to mount the fiber. A tungsten tip with a tip of 1 μm diameter connected to a high precision micro-manipulator was used to manipulate the silica nanofiber sample from the PDMS to the mechanical test platform. Cyanoacrylate glue was used to attach the silica nanofiber to the grips.

Stress and strain were obtained from the force normalized over the cross-section diameter and elongation over the gauge length. Figure 11 shows the stress–strain curve of a silica nanofiber of diameter 108 nm. The Young's modulus was obtained from the slope of the stress–strain curve in the elastic region. For that fiber, it was found to be 67.7 GPa. The strength was 4.5 GPa and its elongation at failure was 9%. A second silica nanofiber was also tested with diameter D = 90 nm (Figure 11). The Young's modulus value was 76.5 GPa, while the ultimate strength was 8.2 GPa, twice that of the previous sample. Elongation at failure was also observed to be higher than the previous sample at 13.8%.

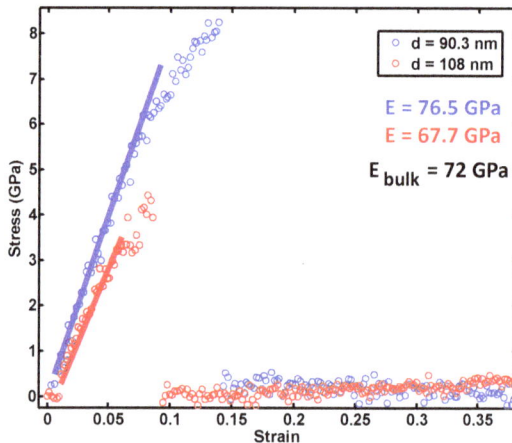

Figure 11. Stress–strain curves for the silica nanofibers.

The Young's modulus values are similar to that of bulk silica (72 GPa) [35]. Silva et al. tested silica nanofibers using an AFM cantilever bending test and found that silica nanofiber Young's modulus for diameters larger than 280 nm was similar to bulk [35]. Wang et al. [36] obtained 27 ± 7 GPa for diameters between 43 nm and 95 nm and Dickin et al. [37] obtained 47 ± 7 GPa for diameters between 80 nm and 98 nm. These measurements were done in the TEM using a resonance method. Recent work [38] on a tensile test of silica nanofibers of dimensions 33.9 nm and 18.7 nm under TEM found that Young's modulus was about 77.6 GPa, close to bulk silica. The values in Figure 11 are in good agreement with the bulk value and with much of the literature.

The strengths are much higher than bulk silica (0.2 GPa) [39]. This is because of nanoscale size effect, where flaw density is significantly reduced and the probability of having a flaw larger than critical flaw is minimal. However, our values of silica nanofiber strength are somewhat lower than that of reference [39], where an average strength of 10 GPa was measured. Our fabrication method of using butane torch and strength test in high relative humidity (~70%) introduces hydroxyl groups in the silica surface that may degrade the silica sample [39,40].

We observed elongation as high as 13.8% in the silica nanofibers. However, SEM imaging of the fractured glass fiber shows a flat fracture surface, consistent with brittle failure. It is likely that the glue slipped to a small degree as failure was approached, as the curve deviated from linearity. However, this only affects the penultimate portion of the stress–strain curve. From this data, it is apparent that the test platform can measure strength and strain if the specimen is firmly gripped.

3.2. Gripping the UHMWPE Nanofibers in Mechanical Testing

For UHMWPE nanofibers, initial test results revealed more severe slip at the grips before their strength was reached. The surface energy of PE is low (31 mJ m^{-2}) and the absence of a polar component (similar to PTFE, 19 mJ m^{-2}) makes it difficult to wet with adhesives [41,42]. There are many commercial including Loctite 3032, Loctite 3035, Scotch weld DP 8005, Scotch weld DP 8010, Loctite plastic bonding system, and TAP poly-weld. They are designed for low surface energy olefins such as polyethylene and polypropylene. However, their viscosity, working life, and/or requirement to surface treat the sample made them unsuitable for this work. All available glues were tried. All slipped to some degree.

Six ultra-drawn NFs with diameters ranging from 85 nm to 255 nm were tested with Loctite® AA 3032/Henkel, which is a specialized glue for PE and was found to be the best glue for this work. From composites theory [43], the maximum force (F_{max}) applied to a specimen of diameter D before it slips can be used to obtain the average interface shear strength, τ_{max}. Accordingly,

$$\tau_{max} = \frac{F_{max}}{\pi D l_{glue}} \tag{1a}$$

which can be rewritten as

$$\frac{F_{max}}{\pi l_{glue}} = \tau_{max} D \tag{1b}$$

Here, l_{glue} is the length of the nanofiber/glue interface. In Figure 12a, $F_{max}/\pi l_{glue}$ is plotted against D. The slope of specimens that slipped before failure, as indicated by the red circles, is 4.7 MPa. This corresponds to τ_{max}. Indeed, video images taken during the testing revealed the slip directly. The τ_{max} value compares reasonably well with the shear strength of Loctite® AA 3032/Henkel, (6 MPa with high density PE). On the other hand, specimens tested until failure (blue circles in Figure 12a) displayed a significantly higher force per unit length. The additional gripping force was contributed by a mechanical locking procedure, as described in [3]. Figure 12b,c shows the schematic of fiber within the glue.

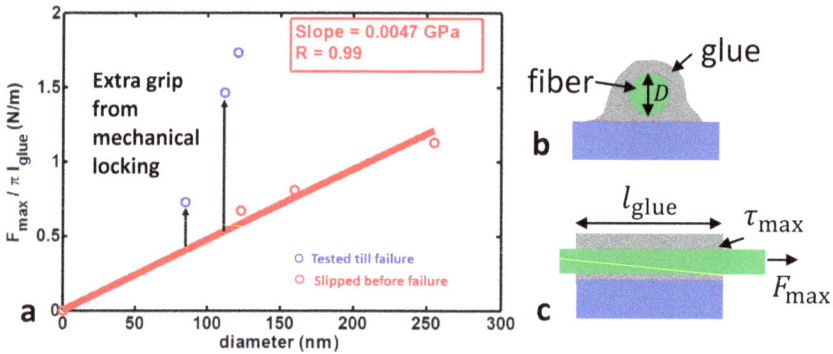

Figure 12. Estimate of glue shear strength. (**a**) Shear force per unit length versus fiber diameter for Loctite® AA 3032/Henkel glue/fiber interface. The slope gives the shear strength of the glue/nanofiber interface. (**b,c**) are schematic cross-section and longitudinal views of the fiber in the glue. Adapted from reference [3] supplementary information.

From this data, we can estimate the length of the glued/fiber interface required to grip a typical 100 nm NF at a uniaxial stress of 10 GPa, on par with the tensile strength, σ_{ts}. Then, $l_{glue} = \sigma_{ts}D/4\tau_{max} \approx 53\ \mu m$. The gripper length was 30 μm. With greater lengths, it should be possible to test UHMWPE nanofiber specimens to failure without the mechanical locking procedure, and thereby also assess Young's modulus.

3.3. Thermal Contact Resistance

Thermal contact resistance R_c between the sample and the islands will result in an incorrect thermal conductivity value if not correctly assessed. Furthermore, if R_c is much larger than the intrinsic thermal resistance of nanofiber, R_s, the procedure is subject to large errors. The total thermal resistance at steady state is the sum of thermal contact resistances, $2R_c$, and R_s. Then,

$$R_{tot} = R_s + 2R_c \qquad (2)$$

For high thermal conductivity samples, such as UHMWPE nanofibers, R_s can be increased by making the nanofiber long and the cross-section area small. However, this will also reduce the heat flux significantly. Consequently, the temperature rise of sensing side will be lessened, making accurate measurement difficult. A platinum or graphite coating, using electron beam or focused ion beam (FIB), has been extensively used in literature to reduce thermal contact resistance [44,45]. However, high-energy electron/ion beam amorphized the NFs and reduced the thermal conductivity down to the bulk value, as shown in Figure 13. There, R_{tot} data is shown over a wide temperature range, but the extracted thermal conductivities of 0.1 to 0.4 W/mK are very low. A better approach is to obtain a condition in which $2R_c \ll R_s$. Values from 60–90 W/mK [3], on par with metals, were measured for NFs attached by the capillary-assisted adhesion method described in Section 2.2.2.

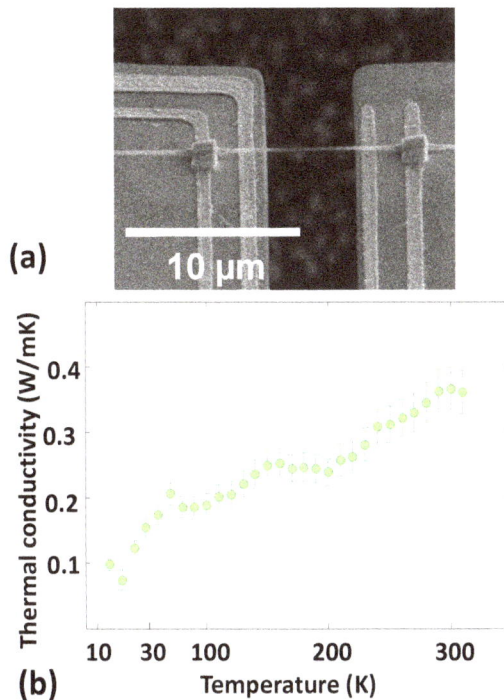

Figure 13. (**a**) NF attached by a platinum coating using FIB. (**b**) The thermal conductivity values of NF prepared by this procedure are similar to bulk PE due to the damage in the NF chains. Adapted from reference [3] supplementary information.

4. Summary and Conclusions

Polymer nanofibers may possess properties much higher than their bulk counterparts, and therefore accurate assessment of these properties is of great interest. However, because of their small size and the omnipresence of van der Waals, electrostatic and capillary forces, it is often difficult to manipulate individual polymer nanofibers on to test platforms that measure their properties. Here, we have detailed the fabrication (Appendix A) and use (Section 2.2.2) of a specimen collector (Figures 1, 2 and 5), which makes it possible to position nanofibers within ±1 μm of the desired location (Figures 6–8) without exposing them to damaging electron beam irradiation. Even after they are in place, the very high strength and low surface energy of UHWMPE nanofiber render grip strength an obstacle to accurate mechanical testing results (Figure 12). This issue can be surmounted by gripping outside the smallest nanofiber region if extra material has accumulated there. Alternatively, longer grip regions may be effective. With respect to thermal measurements, contact resistance can dominate the values (Figure 13). We have found that a capillary-assisted technique (Figure 9) can significantly reduce contact resistance without damage to the nanofiber. While we find these procedures to be helpful, further manipulation improvements will surely be welcomed by practitioners.

Author Contributions: Conceptualization, all the authors; Investigation, R.S.; Writing—Original Draft Preparation, M.P.d.B.; Writing—Review & Editing, R.S. and M.P.d.B.; Supervision, S.S. and M.P.d.B.

Funding: This work was supported by the US National Science Foundation (NSF) under award CMMI-1334630.

Acknowledgments: We thank Kedar Hippalgaonkar (Institute of Materials Research and Engineering (IMRE), Singapore) for providing us with thermal platform.

Conflicts of Interest: The authors declare no conflict of interest.

Appendix A. Specimen Collector Fabrication

The specimen collector was fabricated by bulk micromachining techniques, as shown in Figure A1. The selectivity of a traditional oxide mask for silicon etching is insufficient for a through etch. Four-inch (100)-oriented silicon wafers coated with a 300 nm LPCVD silicon nitride (SiN_x) film on both sides were purchased from WRS materials. The wafers were HMDS vapor-primed at 150 °C for 5 min to promote photoresist adhesion. AZ 4210 photoresist was spin-coated at 4000 rpm, resulting in a 2.1 μm thick layer. This was followed by a soft bake at 95 °C for 1 min.

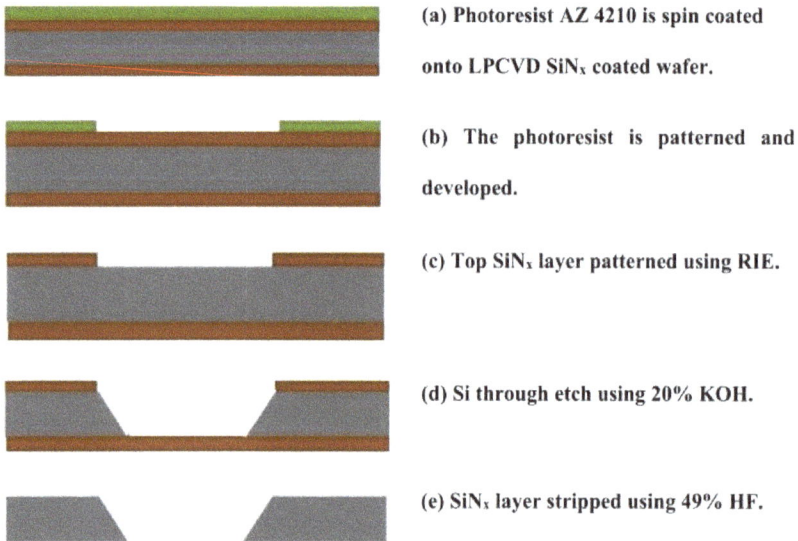

(a) Photoresist AZ 4210 is spin coated onto LPCVD SiN_x coated wafer.

(b) The photoresist is patterned and developed.

(c) Top SiN_x layer patterned using RIE.

(d) Si through etch using 20% KOH.

(e) SiN_x layer stripped using 49% HF.

Figure A1. Micro fabrication processes of the sample collector.

The photoresist was patterned using a Karl Suss MA6 aligner with ultraviolet light of 320 nm wavelength at 5 mW/cm^2 for 80 s. The exposed photoresist was then developed by a 1:4 mixture of AZ 400K and deionized (DI) water for 120 s. The wafer was then cleaned in DI water and dried with nitrogen.

The photoresist pattern was transferred into the SiN_x by reactive ion etching (RIE) using a Plasmatherm 790 RIE system. A gas mixture of tetrafluoromethane (CF_4) and oxygen (O_2) at flow rates of 19.5 sccm and 0.5 sccm, respectively and a total pressure of 35 mT. An RF power of 60 watts at a bias of 315 volts was applied to induce plasma. The fluorine in CF_4 reacts with the silicon in silicon nitride to form volatile silicon tetrafluoride (SiF_4). Carbon can form polymer deposit onto the surface, reducing the etch rate. Hence, the fluorine/carbon ratio determines the Si_xN_y etch rate. A higher ratio provides a higher etch rate. Oxygen can react with the carbon forming volatile gas, which increases the fluorine to carbon ratio and thus increase the reaction rate. However, oxygen also etches the photoresist, so an optimum ratio must be found. We obtained an average etch rate of 26 nm/min but the etch rate decreased over the time due to the polymer deposit. After 6 min, the etching was stopped and argon gas was used to purge the chamber for 2 min at 200 mT. This was done to eliminate polymer deposit that would otherwise hinder the SiN etching. The total duration was 12.5 min to ensure all silicon nitride had been etched. Oxygen plasma was then applied for 5 min at 50 mT and 100 W to ash the photoresist and remove any residual polymer. The chamber was once again purged with argon at 200 mT for 2 min. The wafer samples were then cleaned using acetone, isopropanol alcohol and DI water, respectively.

The wafer was now through-etched using KOH according to SiN_x. pattern. KOH etching depends on the crystal orientation of silicon, the solution concentration and the temperature. Here, KOH pellets were mixed with DI water to obtain 20% solution, an optimum concentration for (100) Si etching [46]. The solution was heated to 80 °C and the wafer was placed in the solution, whereupon hydrogen bubbles, indicative of the etching, were immediately observed. The glassware was covered with aluminum foil to reduce evaporative loss, thereby reducing the KOH concentration increase as the vapor re-condensed into the solution. However, the foil reacted with the vapor and some holes were produced. The total etch duration was 7 h for an average etch rate of 78.5 μm/h, less than the reported etch rate for 20% KOH at 80 °C [46], likely due to the increasing concentration. Etching ceased at the bottom SiN_x layer. KOH etching is an anisotropic etch with the (111) plane etching more slowly than the (110) and (100) planes, resulting in a characteristic pyramid like structure with 54.7° [46]. At the end of KOH etch, we obtained square holes with the expected 54.7° slanted walls. The wafer was rinsed in DI water, and the remaining SiN_x material on the top and bottom was removed after 15 min in 49% hydrofluoric acid (HF). The wafer was then diced into individual specimen collectors.

Appendix B. Microheater Construction

The micro heater is a sharply bent tungsten wire locally etched at the tip to 5–10 μm. It is joule-heated to a temperature exceeding 500 °C (Figure 3) at its tip to locally transform a MF into a NF. It was constructed and assembled as follows: Tungsten wire of 50 μm diameter was purchased from ESPI metals, cut to a 10 cm length, bent near its center, and adhered to a glass slide. The protruding bent edge was submerged in a beaker with 30% H_2O_2 solution that was held at room temperature while it was stirred magnetically. The wire was etched until the final diameter at the tip reached 5–10 μm as shown in Figure A2a. The ends of the wire were attached with copper tape for electrical attachment. Figure A2b shows a fabricated microheater.

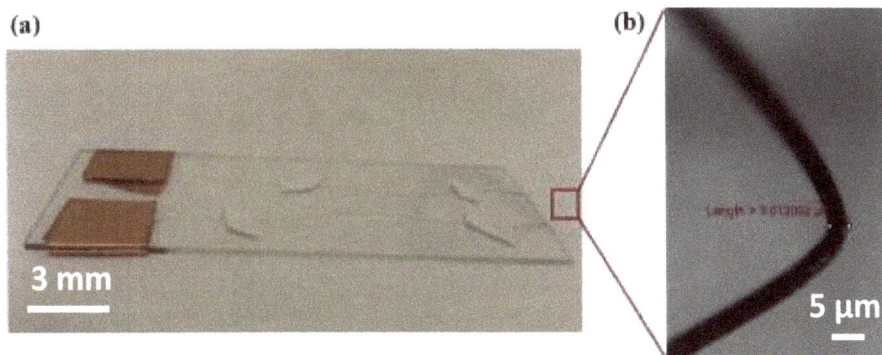

(a)

3 mm

(b)

5 μm

Figure A2. Microheater used for local heat stretching and cutting PE fiber. (**a**) microheater (**b**) tungsten tip of microheater with diameter ~3 μm.

References

1. Naraghi, M.; Chasiotis, I.; Kahn, H.; Wen, Y.; Dzenis, Y. Novel method for mechanical characterization of polymeric nanofibers. *Rev. Sci. Instrum.* **2007**, *78*, 85108. [CrossRef] [PubMed]
2. Li, P.; Hu, L.; McGaughey, A.J.H.; Shen, S. Crystalline polyethylene nanofibers with the theoretical limit of Young's modulus. *Adv. Mater.* **2014**, *26*, 1065–1070. [CrossRef] [PubMed]

3. Shrestha, R.; Li, P.; Chatterjee, B.; Zheng, T.; Wu, X.; Liu, Z.; Luo, T.; Choi, S.; Hippalgaonkar, K.; de Boer, M.P.; et al. Crystalline polymer nanofibers with ultra-high strength and thermal conductivity. *Nat. Commun.* **2018**, *9*, 1664. [CrossRef] [PubMed]

4. Shen, S.; Henry, A.; Tong, J.; Zheng, R.; Chen, G. Polyethylene nanofibres with very high thermal conductivities. *Nat. Nanotechnol.* **2010**, *5*, 251–255. [CrossRef] [PubMed]

5. Zhong, Z.; Wingert, M.C.; Strzalka, J.; Wang, H.H.; Sun, T.; Wang, J.; Chen, R.; Jiang, Z. Structure-induced enhancement of thermal conductivities in electrospun polymer nanofibers. *Nanoscale* **2014**, *6*, 8283–8291. [CrossRef] [PubMed]

6. Singh, V.; Bougher, T.L.; Weathers, A.; Cai, Y.; Bi, K.; Pettes, M.T.; McMenamin, S.A.; Lv, W.; Resler, D.P.; Gattuso, T.R.; et al. High thermal conductivity of chain-oriented amorphous polythiophene. *Nat. Nanotechnol.* **2014**, *9*, 384–390. [CrossRef] [PubMed]

7. Agarwal, S.; Wendorff, J.H.; Greiner, A. Progress in the Field of Electrospinning for Tissue Engineering Applications. *Adv. Mater.* **2009**, *21*, 3343–3351. [CrossRef] [PubMed]

8. Chronakis, I.S. Novel nanocomposites and nanoceramics based on polymer nanofibers using electrospinning process—A review. *J. Mater. Process. Technol.* **2005**, *167*, 283–293. [CrossRef]

9. Huang, Z.M.; Zhang, Y.Z.; Kotaki, M.; Ramakrishna, S. A review on polymer nanofibers by electrospinning and their applications in nanocomposites. *Compos. Sci. Technol.* **2003**, *63*, 2223–2253. [CrossRef]

10. Fang, J.; Niu, H.; Lin, T.; Wang, X. Applications of electrospun nanofibers. *Sci. Bull.* **2008**, *53*, 2265–2286. [CrossRef]

11. Calderón, M.Á.R.; Zhao, W. Applications of Polymer Nanofibers in Bio-Materials, Biotechnology and Biomedicine: A Review. In *TMS 2014: 143rd Annual Meeting & Exhibition*; Springer International Publishing: Cham, Switzerland, 2014; pp. 401–414.

12. Fu, Y.Q.; Luo, J.K.; Flewitt, A.J.; Milne, W.I. Smart microgrippers for bioMEMS applications. *MEMS Biomed. Appl.* **2012**, 291–336. [CrossRef]

13. Bøggild, P. Nanogrippers. In *Encyclopedia of Nanotechnology*; Bhushan, B., Ed.; Springer: New York, NY, USA, 2015.

14. Agnus, J.; Nectoux, P.; Chaillet, N. Overview of microgrippers and design of a micro-manipulation station based on a MMOC microgripper. In Proceedings of the 2005 IEEE International Symposium on Computational Intelligence in Robotics and Automation, Espoo, Finland, 27–30 June 2005.

15. Soltani, M.; Lin, J.; Forties, R.A.; Inman, J.T.; Saraf, S.N.; Fulbright, R.M.; Lipson, M.; Wang, M.D. Nanophotonic trapping for precise manipulation of biomolecular arrays. *Nat. Nanotechnol.* **2014**, *9*, 448–452. [CrossRef] [PubMed]

16. Nam, A.S.; Amon, C.; Sitti, M. Three-dimensional nanoscale manipulation and manufacturing using proximal probes: Controlled pulling of polymer micro/nanofibers. In Proceedings of the IEEE International Conference on Mechatronics (ICM'04), Istanbul, Turkey, 5 June 2004; pp. 224–230.

17. Wingert, M.C.; Jiang, Z.; Chen, R.; Cai, S. Strong size-dependent stress relaxation in electrospun polymer nanofibers. *J. Appl. Phys.* **2017**, *121*, 015103. [CrossRef]

18. Egerton, R.F.; Li, P.; Malac, M. Radiation damage in the TEM and SEM. *Micron* **2004**, *35*, 399–409. [CrossRef] [PubMed]

19. Doshi, J.; Reneker, D.H. Electrospinning process and applications of electrospun fibers. *J. Electrostat.* **1995**, *35*, 151–160. [CrossRef]

20. Reneker, D.H.; Chun, I. Nanometre diameter fibres of polymer, produced by electrospinning. *Nanotechnology* **1996**, *7*, 216–223. [CrossRef]

21. Shi, X.M.; Bin, Y.Z.; Hou, D.S.; Men, Y.F.; Matsuo, M. Gelation/crystallization mechanisms of UHMWPE solutions and structures of ultradrawn gel films. *Polym. J.* **2014**, *46*, 21–35. [CrossRef]

22. Features of ANSYS Fluent. Available online: http://www.ansys.com/products/fluids/ansys-fluent/ansys-fluent-features (accessed on 14 January 2018).

23. Narayanaswamy, A.; Gu, N. Heat Transfer from Freely Suspended Bimaterial Microcantilevers. *J. Heat Transf.* **2011**, *133*, 042401. [CrossRef]

24. DSM Dyneema Fact Sheet. Available online: http://www.pelicanrope.com/pdfs/Dyneema-Comprehensive-factsheet-UHMWPE.pdf (accessed on 14 January 2018).

25. Fujikura, Y.; Suzuki, T.; Matsumoto, M. Emissivity of chlorinated polyethylene. *J. Appl. Polym. Sci.* **1982**, *27*, 1293–1300. [CrossRef]

26. Shroff, S.S.; de Boer, M.P. Constant Velocity High Force Microactuator for Stick-Slip Testing of Micromachined Interfaces. *J. Microelectromech. Syst.* **2015**, *24*, 1868–1877. [CrossRef]

27. Sniegowski, J.J.; de Boer, M.P. IC-Compatible Polysilicon Surface Micromachining. *Annu. Rev. Mater. Sci.* **2000**, *30*, 299–333. [CrossRef]

28. Weibel, G.L.; Ober, C.K. An overview of supercritical CO_2 applications in microelectronics processing. *Microelectron. Eng.* **2003**, *65*, 145–152. [CrossRef]

29. Sawyer, L.C.; Grubb, D.T.; Meyers, G.F. *Polymer Microscopy*; Springer: New York, NY, USA, 2008; ISBN 978-0-387-72627-4.

30. Revol, J.F.; Manley, R.S.J. Lattice imaging in polyethylene single crystals. *J. Mater. Sci. Lett.* **1986**, *5*, 249–251. [CrossRef]

31. Ma, J.; Zhang, Q.; Mayo, A.; Ni, Z.; Yi, H.; Chen, Y.; Mu, R.; Bellan, L.M.; Li, D. Thermal conductivity of electrospun polyethylene nanofibers. *Nanoscale* **2015**, *7*, 16899–16908. [CrossRef] [PubMed]

32. Naraghi, M.; Ozkan, T.; Chasiotis, I.; Hazra, S.S.; de Boer, M.P. MEMS platform for on-chip nanomechanical experiments with strong and highly ductile nanofibers. *J. Micromech. Microeng.* **2010**, *20*, 125022. [CrossRef]

33. Li, D.; Wu, Y.; Kim, P.; Shi, L.; Yang, P.; Majumdar, A. Thermal conductivity of individual silicon nanowires. *Appl. Phys. Lett.* **2003**, *83*, 2934–2936. [CrossRef]

34. Tong, L.; Gattass, R.R.; Ashcom, J.B.; He, S.; Lou, J.; Shen, M.; Maxwell, I.; Mazur, E. Subwavelength-diameter silica wires for low-loss optical wave guiding. *Nature* **2003**, *426*, 816–819. [CrossRef] [PubMed]

35. Silva, E.C.C.M.; Tong, L.; Yip, S.; Van Vliet, K.J. Size Effects on the Stiffness of Silica Nanowires. *Small* **2006**, *2*, 239–243. [CrossRef] [PubMed]

36. Wang, Z.; Gao, R.; Poncharal, P.; De Heer, W.; Dai, Z.; Pan, Z. Mechanical and electrostatic properties of carbon nanotubes and nanowires. *Mater. Sci. Eng. C* **2001**, *16*, 3–10. [CrossRef]

37. Dikin, D.A.; Chen, X.; Ding, W.; Wagner, G.; Ruoff, R.S. Resonance vibration of amorphous SiO_2 nanowires driven by mechanical or electrical field excitation. *J. Appl. Phys.* **2003**, *93*, 226–230. [CrossRef]

38. Luo, J.; Wang, J.; Bitzek, E.; Huang, J.Y.; Zheng, H.; Tong, L.; Yang, Q.; Li, J.; Mao, S.X. Size-Dependent Brittle-to-Ductile Transition in Silica Glass Nanofibers. *Nano Lett.* **2016**, *16*, 105–113. [CrossRef] [PubMed]

39. Brambilla, G.; Payne, D.N. The Ultimate Strength of Glass Silica Nanowires. *Nano Lett.* **2009**, *9*, 831–835. [CrossRef] [PubMed]

40. Armstrong, J.L.; Matthewson, M.J.; Kurkjian, C.R. Humidity Dependence of the Fatigue of High-Strength Fused Silica Optical Fibers. *J. Am. Ceram. Soc.* **2000**, *83*, 3100–3109. [CrossRef]

41. Owens, D.K.; Wendt, R.C. Estimation of the surface free energy of polymers. *J. Appl. Polym. Sci.* **1969**, *13*, 1741–1747. [CrossRef]

42. Brewis, D.M.; Briggs, D. Adhesion to polyethylene and polypropylene. *Polymer* **1981**, *22*, 7–16. [CrossRef]

43. Chawla, K.K. Ceramic Matrix Composites. In *Composite Materials*; Springer: New York, NY, USA, 1998; pp. 212–251.

44. Yu, C.; Saha, S.; Zhou, J.; Shi, L.; Cassell, A.M.; Cruden, B.A.; Ngo, Q.; Li, J. Thermal Contact Resistance and Thermal Conductivity of a Carbon Nanofiber. *J. Heat Transf.* **2006**, *128*, 234–239. [CrossRef]

45. Cheng, Z.; Liu, L.; Xu, S.; Lu, M.; Wang, X. Temperature Dependence of Electrical and Thermal Conduction in Single Silver Nanowire. *Sci. Rep.* **2015**, *5*, 10718. [CrossRef] [PubMed]

46. Seidel, H.; Csepregi, L.; Heuberger, A.; Baumgärtel, H. Anisotropic Etching of Crystalline Silicon in Alkaline Solutions. *J. Electrochem. Soc.* **1990**, *137*, 3612–3626. [CrossRef]

Article

An Approach to the Extreme Miniaturization of Rotary Comb Drives

Andrea Veroli [1,*], Alessio Buzzin [1], Fabrizio Frezza [1], Giampiero de Cesare [1], Muhammad Hamidullah [1,2], Ennio Giovine [2], Matteo Verotti [3] and Nicola Pio Belfiore [4]

[1] Department of Information Engineering, Electronics and Telecommunications, University of Rome La Sapienza, 00156 Rome, Italy; alessio.buzzin@uniroma1.it (A.B.); fabrizio.frezza@uniroma1.it (F.F.); giampiero.decesare@uniroma1.it (G.d.C.)

[2] Institute of Photonics and Nanotechnologies, IFN-CNR, Via Cineto Romano 42, 00156 Rome, Italy; m.hamidullah@ifn.cnr.it (M.H.); ennio.giovine@cnr.it (E.G.)

[3] University of Rome Niccolò Cusano, 00156 Rome, Italy; matteo.verotti@uniroma1.it

[4] Department of Engineering, University of Roma Tre, 00156 Rome, Italy; nicolapio.belfiore@uniroma3.it

* Correspondence: andrea.veroli@uniroma1.it; Tel.: +39-06-44585-379

Received: 7 September 2018; Accepted: 6 October 2018; Published: 11 October 2018

Abstract: The evolution of microelectronic technologies is giving constant impulse to advanced micro-scaled systems which perform complex operations. In fact, the actual micro and nano Electro-Mechanical Systems (MEMS/NEMS) easily integrate information-gathering and decision-making electronics together with all sorts of sensors and actuators. Mechanical manipulation can be obtained through microactuators, taking advantage of magnetostrictive, thermal, piezoelectric or electrostatic forces. Electrostatic actuation, more precisely the comb-drive approach, is often employed due to its high versatility and low power consumption. Moreover, the device design and fabrication process flow can be simplified by compliant mechanisms, avoiding complex elements and unorthodox materials. A nano-scaled rotary comb drive is herein introduced and obtained using NEMS technology, with an innovative design which takes advantages of the compliant mechanism characteristics. A theoretical and numerical study is also introduced to inspect the electro-mechanical behavior of the device and to describe a new technological procedure for its fabrication.

Keywords: microactuation; nanoactuation; comb drives; flexure hinge; nanofabrication; electron beam lithography

1. Introduction

The end of the 1960s has witnessed the rise and consolidation of semiconductor fabrication technologies leading, together with the first microchip developments, to the evolution of increasingly complex systems able to connect advanced electronics and engineering systems with the physical world [1]. This process was made possible due to the development of MEMS sensors and actuators, that are currently an essential part of our daily life, being present in information-communication, electromechanical, optical, chemical and biological devices [2–5]. Moreover, the scaling phenomenon from macro- to micro-actuators led to a drastic change in the influence of individual parameters and opened new perspectives with innovative mechanical designs.

MEMS actuators and sensors are many and different. Micromachined structures movimentation can be carried out using various approaches: such as magnetostrictive, thermal, piezoelectric or electrostatic [6]. In this work, attention is given to work-producing actuators, with silicon being the key material, by an electrical and mechanical point of view. State-of-the-art MEMS devices are often actuated taking advantage of electrostatic forces between charged surfaces at a distance. Electrostatic actuation combines high versatility, fast dynamic response and low power consumption with a simplified design and a fabrication process flow which is made possible by avoiding complex elements like coils or cores and unorthodox materials like shape-memory-alloys or piezoelectric ceramics [7]. It exploits the relation of surface to spacing (not volume to spacing), therefore its force is less affected by scaling [8].

Among all the possible electrostatic actuation techniques, the comb-drive approach is one of the most employed [9], with different possibilities in terms of out-of-plane [10] or in-plane [11] motion.

Moreover, Compliant mechanisms have been successfully adopted to develop MEMS devices in order to replace hinges and improve motion properties: the incorporation of the Conjugate Surfaces Flexure Hinges (CSFH) mechanism into the MEMS technology has opened up new perspectives for a completely new class of microdevices, characterized by a neutral stable configuration with a stationary pose, avoiding lubrication due to the absence of mechanical backlash, and easier to actuate [12]. The outcome is not a combination of different layers and sub-parts, the whole device can be made of a monolithic body of a single material, obtained by a highly simplified fabrication procedure, being at the same time extremely versatile in terms of integration in complex systems and structures for different applications, such as compliant 3 D.O.F. microrobots [13–16], micromechanisms [17,18], micro hinges [19–21] and microgrippers [22–27].

The drastic reduction of complexity made possible by the CSFH principle, together with its minimization of internal stresses and its robustness in operation, can facilitate the scaling process from a Micro-Electro-Mechanical System to a Nano-Electro-Mechanical System (NEMS), with new challenges in terms of mechanical movement, actuation possibilities and applications [28–30]. In this work, the properties of a NEMS rotary comb-drive nano-scaled actuator is introduced, its design is inspected, its electro-mechanical behaviour is theoretically and numerically studied and a new technological procedure for its fabrication is described, making it able to integrate in standard systems as well as in wearable/flexible systems, polymer substrates or in glass devices, with a variety of applications in the field of mechanical manipulation, such as nano-surgery, lab-on-chip bimolecular analysis, or nano-movimentation in space environment.

2. A Nano-Scaled Rotary Comb-Drive Electrostatic Actuator

The presented structure covers a total area of $120 \times 75\ \mu m^2$ and is made of 2 main components, as depicted in Figure 1:

- A fixed part (red colored), with interdigitated fingers;
- A suspended part (green colored), composed by a mobile rigid body with interdigitated fingers and a flexure hinge with an anchored part on the left.

The system can be moved by changing the voltage between the static and movable part, which generates an alteration in the electrostatic forces between the two components, which are proportional to the capacitance between the two interdigitated structures: the result is a variation of the gap between the comb fingers and a movement of the rigid suspended body, made possible by the flexible hinge. Figure 2 shows a 3D sketch of the system.

The actual comb-drive mechanism is $50 \times 25\ \mu m^2$, the smallest size ever presented in literature (to the best of the authors' knowledge), and has two arrays of 18,600 nm-wide and 1800 nm-spaced fingers, with a length which changes from 4 μm up to 9 μm; the first and last finger of each array is 1200 nm-wide. The whole suspended part is able to move due to a 600 nm wide curved beam flexure hinge, with a circular shape and a radius of 20 μm. The curved beam is considered as a revolute joint, centered in correspondence of the center of the beam elastic weight.

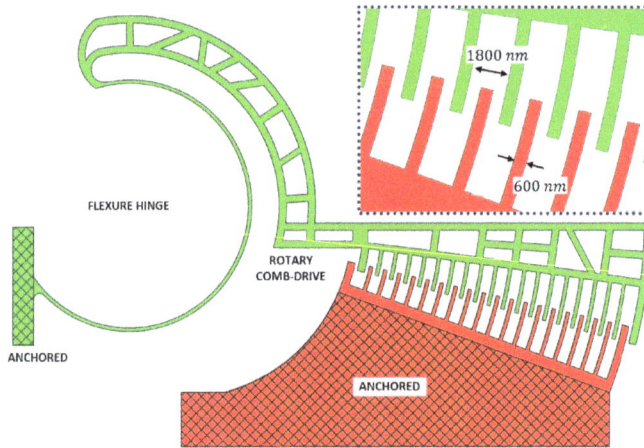

Figure 1. Top view of the proposed device.

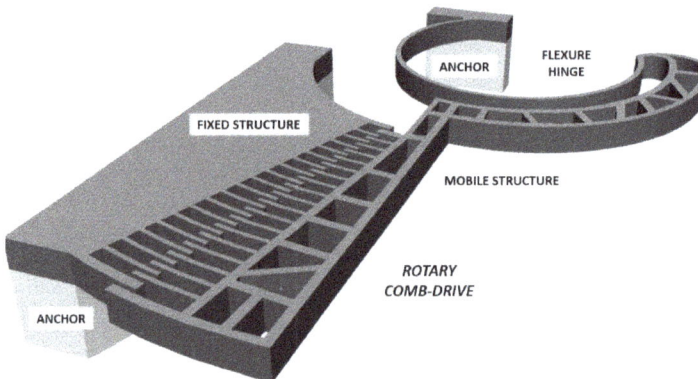

Figure 2. Three-D sketch of the proposed device.

3. Microelectromechanical Simulation

The response of the comb drive to the application of a voltage on the pads connected to the fixed and mobile sets of fingers has been simulated in two steps.

Firstly, considering the geometry of the systems, the number of fingers and the dielectric characteristics of the gap, an elementary force F has been calculated as acting on each one of the n fingers. Force F has the line of action perpendicular to the finger free-end section and can be calculated as [31]

$$F = \frac{\epsilon_0 \epsilon_r h V^2}{g} ,\qquad (1)$$

where ϵ_0 is the vacuum permittivity (8.8541×10^{-12} Fm^{-1}), ϵ_r is the air relative permittivity (1.00058), h is the thickness of the finger, g is the radial distance between the movable and fixed finger, and V is the applied voltage.

Then, a static force analysis has been performed by means of the commercial Finite Elements Analysis software ANSYS$^{©}$ (Canonsburg, PA, USA). Polycrystalline silicon has been considered as structural material, with Young's modulus equal to 164 GPa and Poisson's ratio equal to 0.22 [32].

Isotropic material formulation and nonlinearity due to large deflections were also considered. A fixed support was introduced to simulate the anchor constraint at the flexure end–section.

Two different layouts for the comb drives have been considered.

3.1. First Layout

For the first layout, $n = 19$ fingers, a thickness $h = 500$ nm and a gap $g = 600$ nm have been considered. The forces were calculated for an applied voltage ranging from 1 to 14 V, with steps of 1 V.

Figure 3 shows the generated mesh, composed of 9484 nodes and 7594 elements, and refined in correspondence to the flexible beam and to the comb-drive fingers.

The rotation of the floating part and the maximum value of the maximum principal stress (MPS), for increasing values of the applied voltage, are reported in Figure 4. The value of 14 V corresponds a rotation of 4.85 °, close to the maximum rotation permitted by the comb-drive geometry (5°). This case is represented in Figure 5, showing the deformed and neutral configurations of the structure and the stress distribution on the flexure. The maximum MPS value is equal to 66.7 MPa. Yield strength of polycrystalline silicon is equal to 1.2 GPa [33,34].

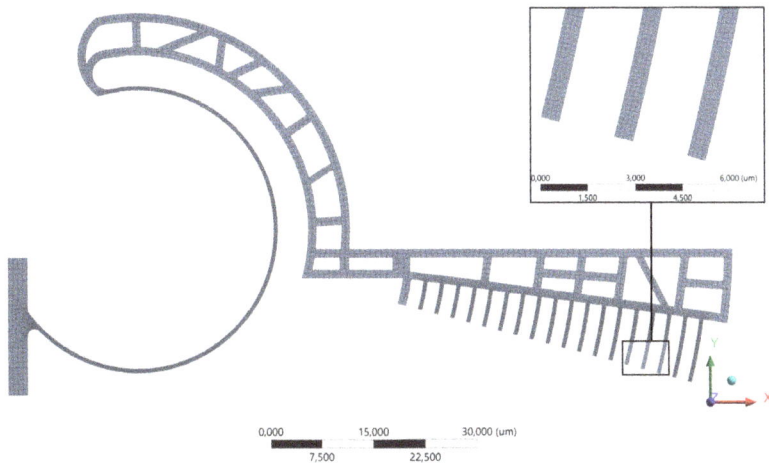

Figure 3. First layout: mesh of the floating structure and detail of the comb-drive fingers.

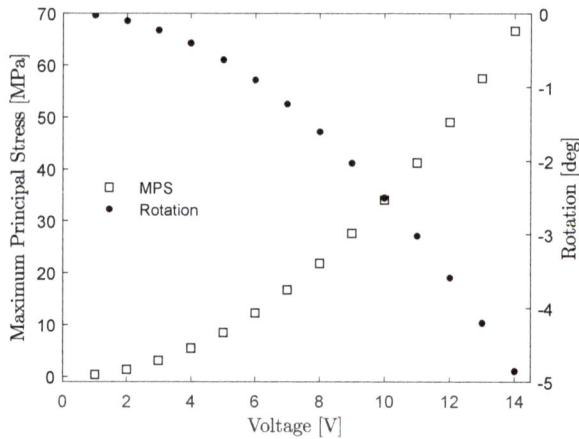

Figure 4. First layout: maximum values of the MPS and rotation angle of the movable fingers for increasing values of the applied voltage V.

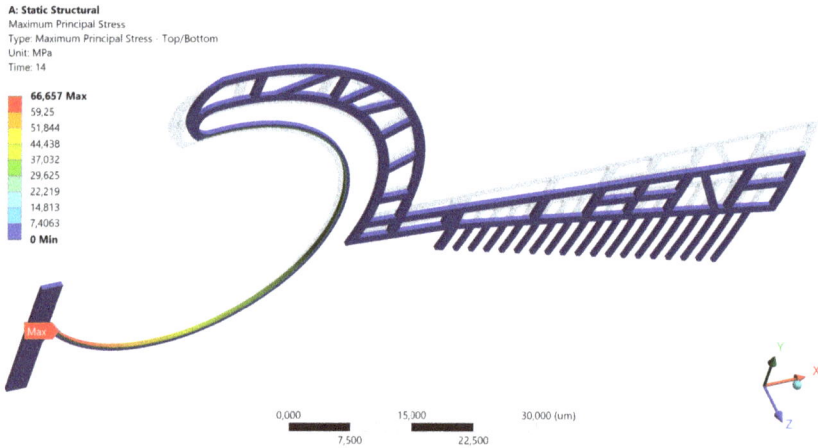

Figure 5. First layout: neutral and deformed configurations, and stress distribution on the flexure.

A modal analysis has been performed with ANSYS to evaluate the dynamic performance of the actuator. The first resonant mode, occurring at 16.5 kHz, corresponds to an out-of-plane motion characterized by an axis of rotation parallel to the y-axis (see Figure 5). This occurrence is not really critical for the fingers which are subject locally to a relative translation with no problems for fingers interference.

The second mode occurs at 18.3 kHz, and it is associated to an in-plane oscillation with an axis of rotation parallel to the z-axis. This mode corresponds, actually, to the designed working motion for the mobile fingers and therefore there are no interference problems.

The third and sixth modes, corresponding to out-of plane motions, are characterized by natural frequencies equal to 41.6 kHz and 149.3 kHz, respectively.

The fourth and fifth modes, occurring at 56.5 kHz and 126.9 kHz, respectively, are attributed to in-plane motions that could be detrimental because of the possible contact among floating and anchored fingers. Finally, higher order modes have been neglected because the natural frequencies are very high.

3.2. Second Layout

After the former simulation, a second case has been studied with a gap g between the fingers larger than in the first case. In fact, the gap has been increased by the 30%, in order to avoid surface sticking during the fabrication and the releasing of the device. Furthermore, this could help operational performances during actuation.

To model the comb drive action in this second layout, the force F has been applied to each one of the $n = 13$ fingers. The thickness of the finger h remained 500 nm, as well as the radial distance between the movable and fixed finger (1.2 µm), while the range of applied voltage spans from 1 to 24 V, with 1 V steps.

Figure 6 reports the generated mesh, composed of 8065 nodes and 6423 elements, and refined in correspondence to the flexible beam and to the comb drive fingers.

The rotation of the floating part and the maximum value of the maximum principal stress (MPS), for increasing values of the applied voltage, are reported in Figure 7. The value of 24 V corresponds to a rotation of 4.88 °, close to the maximum rotation permitted by the comb-drive geometry (5°). As for the previous layout, the limit case is reported in Figure 8, that shows the deformed and neutral configurations and the stress distribution. The maximum MPS value is equal to 66.7 MPa.

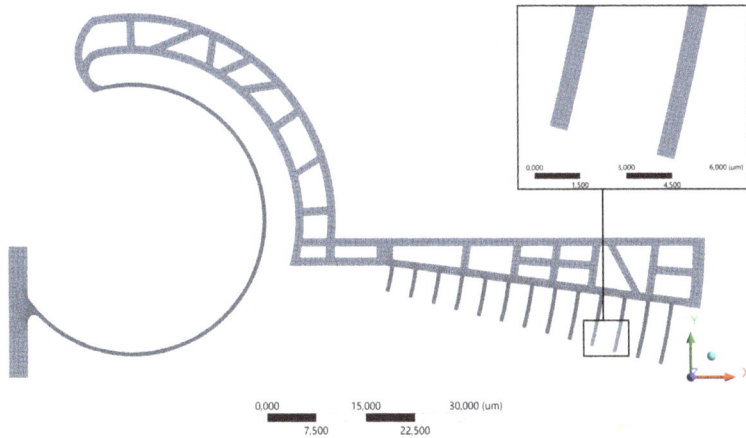

Figure 6. Second layout: mesh of the floating structure and detail of the comb-drive fingers.

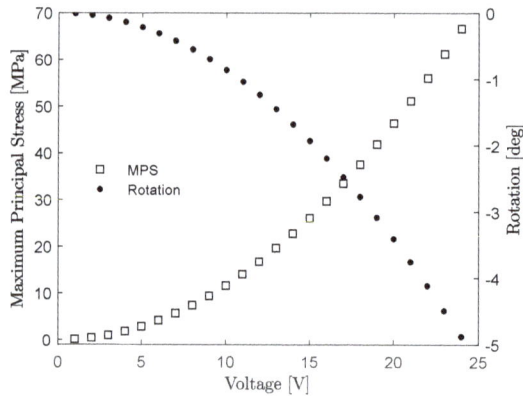

Figure 7. Second layout: maximum values of the MPS and rotation angle of the movable fingers for increasing values of the applied voltage.

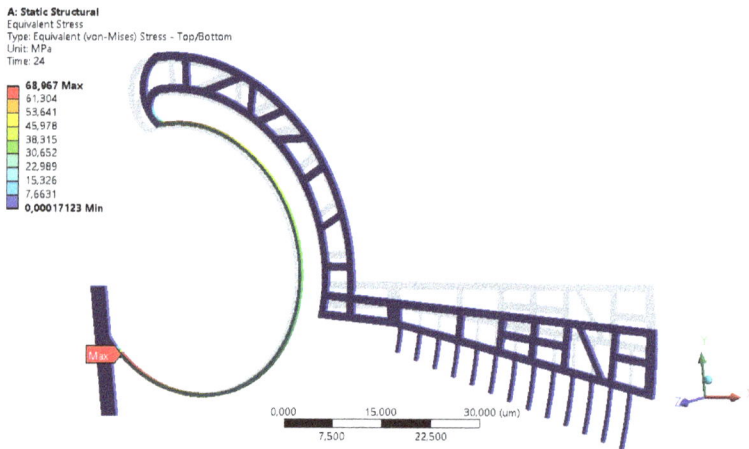

Figure 8. Second layout: neutral and deformed configurations, and stress distribution on the flexure.

As for the first layout, modal analysis has been conducted to calculate the most affecting natural modes and so the first six natural frequencies have been considered, having, respectively, values equal to 16.7, 18.6, 40.4, 57.7, 127.1 and 147.9 kHz. The corresponding modes are similar to the ones described in the previous case.

4. Fabrication

Microelectronic manufacturing technologies, and its constant evolution in time, opened new possibilities for the fabrication of micro- and nano-scaled electro-mechanical systems, allowing to obtain a highly complex mechanism through a sequence of iterations of simple steps:

- Deposition/growth of a chosen material on a substrate in form of thin/thick film, using various kinds of methods, such as Physical or Chemical Vapor Deposition (PVD/CVD), Atomic Layer Deposition (ALD) and Epitaxy;
- Geometry definition of the latter, usually via lithographic techniques or similar;
- Material patterning, via wet/dry etching procedures.

The result is a device assembled by a composition of different layers overlapped, each shaped with the corresponding geometry [35,36].

MEMS manufacturing and production, for this kind of devices, usually employs a standard Silicon-on-Insulator (SOI) stack. For this work, an unconventional material stack is designed, as sketched in Figure 9.

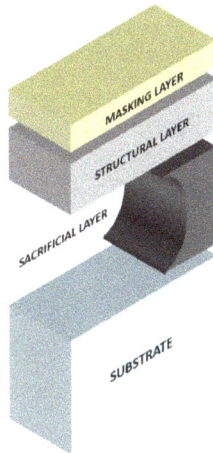

Figure 9. Three-D sketch of the proposed device.

In particular, the stack is composed by:

- A glass substrate;
- A Ti/W metallic alloy as intermediate sacrificial layer;
- Doped hydrogenated amorphous silicon (a-Si:H) as structural layer: a highly versatile material, suitable for large-area and low-cost electronic devices, biologically compatible, while monolithically integrable on rigid or flexible, metallic, dielectric or polymeric substrates [37–42]. It also shows interesting mechanical features and was chosen to build bridge-like structures and cantilevers based devices [43].
- Chromium as masking layer for the structure patterning.

The fabrication procedure was carried out as follows:

- PVD of the sacrificial layer;
- CVD of the amorphous silicon layer;
- Geometry definition on a PVD obtained chromium film by lithographic technique: in particular, Electron Beam Lithography (EBL) was required, in order to obtain submicrometric resolution, on an electron-sensible polymeric film [44];
- Structure patterning by Reactive Ion Etching (RIE) technique, with the previously shaped chromium film as masking layer;
- Structure releasing via isotropic and selective removal of the intermediate layer, in a wet etching solution. Etch holes in the rigid suspended bodies were considered in order to promote the under-etching phenomenon and optimize the releasing process.

Figure 10 shows first examples of the nanoscaled rotary comb-drive during its fabrication process. Since the structure geometries reach the sub-micrometric scale, it was impossible to appropriately identify and discern all the features of our device with an optical microscope: therefore Scanning Electron Microscope (SEM) pictures were taken.

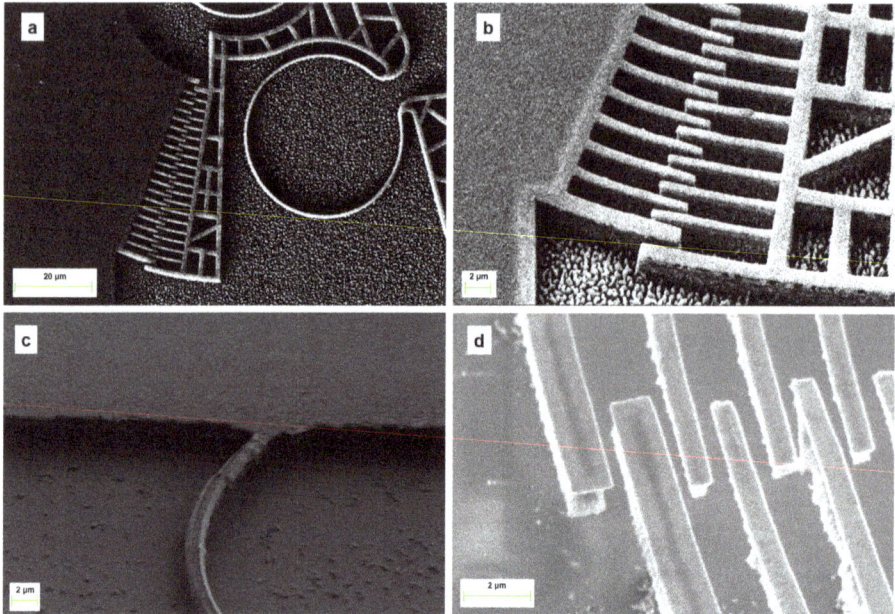

Figure 10. SEM pictures of the nano comb-drive during its fabrication process: top view of the whole patterned device (**a**) with a 30° tilted enlargement on the comb-drive fingers (**b**); 45° tilted view of an anchored flexure hinge (**c**) and fingers (**d**) released from the glass substrate and suspended. These images were taken with an accelerating voltage (EHT) of 10 kV, 10 kV, 2 kV and 5 kV respectively; a magnification of 2630×, 8630×, 8000× and 2281×, respectively; a working distance of 5.5 mm, 10 mm, 5.9 mm and 10.0 mm, respectively, and a beam current of 10 pA.

Figure 10a,b show a top view (top left) of the whole device during the structure patterning phase, with an enlargement on the comb fingers (top right). Figure 10c,d show the crucial parts of the device suspended and detached from the substrate after the releasing process: the attachment between the hinge of the actuator and the anchor (bottom left), and the comb-drive fingers (bottom right). The displayed devices have n-doped a-Si:H, deposited by Plasma Enhanced Chemical Vapour Deposition (PECVD), as structural layer, with the following parameters:

- A substrate temperature of 200 °C;
- A SiH_4 flow of 45 sccm and a PH_3 flow of 10 sccm into the process chamber, with a process pressure of 0.3 Torr;
- A glow discharge with a power density of 25 mW/cm^2, for 1 h.

5. Conclusions

This paper has shown the feasibility of a new nano fabrication process to build nano actuators, based on electrostatic actuation. The construction process has been applied to a rotary comb-drive which appears promising as a tool for microactuation and micromanipulation. Furthermore, the numerical simulations gave also encouraging results in terms of operational capability. Two models have been studied with different characteristics and the results showed that operational capabilities can be increased by reducing the gap between the fingers and their size, which makes the systems less robust. Therefore, an optimal design should find a balance of mediation between the two different factors. The developed actuator seems to be one of the smallest actuators available as state-of-the-art, and this could lead to new applications, especially in the biomedical field.

Author Contributions: A.V., A.B., M.H., and E.G. were involved in the fabrication and design of the rotary comb drives; M.V. and N.P. B. have performed the numerical analysis and design; F.F., G.d.C. and N.P.B coordinated the work and supervised the project.

Conflicts of Interest: The authors declare no conflict of interest.

References

1. Gardner, J.W.; Varadan, V.K.; Awadelkarim, O.O. Microsensors. In *Microsensors, MEMS, and Smart Devices*; John Wiley & Sons, Ltd.: Hoboken, NJ, USA, 2001; pp. 227–302.
2. Bhushan, B. Nanotribology and nanomechanics of MEMS/NEMS and BioMEMS/BioNEMS materials and devices. In *Nanotribology and Nanomechanics*; Springer: Cham, Switzerland, 2017; pp. 797–907.
3. Grayson, A.C.R.; Shawgo, R.S.; Johnson, A.M.; Flynn, N.T.; Li, Y.; Cima, M.J.; Langer, R. A BioMEMS review: MEMS technology for physiologically integrated devices. *Proc. IEEE* **2004**, *92*, 6–21. [CrossRef]
4. Tilmans, H.A.; De Raedt, W.; Beyne, E. MEMS for wireless communications: From RF-MEMS components to RF-MEMS-SiP. *J. Micromech. Microeng.* **2003**, *13*, S139. [CrossRef]
5. Saxena, V.; Plum, T.J.; Jessing, J.R.; Baker, R.J. Design and fabrication of a MEMS capacitive chemical sensor system. In Proceedings of the 2006 IEEE Workshop on Microelectronics and Electron Devices, WMED'06, Boise, ID, USA, 14 April 2006.
6. Bell, D.J.; Lu, T.; Fleck, N.A.; Spearing, S.M. MEMS actuators and sensors: Observations on their performance and selection for purpose. *J. Micromech. Microeng.* **2005**, *15*, S153. [CrossRef]
7. Del Corro, P.G.; Imboden, M.; Pérez, D.J.; Bishop, D.J.; Pastoriza, H. Single ended capacitive self-sensing system for comb drives driven XY nanopositioners. *Sens. Actuators A Phys.* **2018**, *271*, 409–417. [CrossRef]
8. Thielicke, E.; Obermeier, E. Microactuators and their technologies. *Mechatronics* **2000**, *10*, 431–455. [CrossRef]
9. Legtenberg, R.; Groeneveld, A.; Elwenspoek, M. Comb-drive actuators for large displacements. *J. Micromech. Microeng.* **1996**, *6*, 320. [CrossRef]
10. Xie, H.; Pan, Y.; Fedder, G.K. A CMOS-MEMS mirror with curled-hinge comb drives. *J. Microelectromech. Syst.* **2003**, *12*, 450–457.
11. Yi, B.J.; Chung, G.B.; Na, H.Y.; Kim, W.K.; Suh, I.H. Design and experiment of a 3-DOF parallel micromechanism utilizing flexure hinges. *IEEE Trans. Robot. Autom.* **2003**, *19*, 604–612.
12. Belfiore, N.P.; Scaccia, M.; Ianniello, F.; Presta, M. Selective Compliance Hinge. U.S. Patent 8,191,204 B2, 6 May 2012.
13. Balucani, M.; Belfiore, N.P.; Crescenzi, R.; Genua, M.; Verotti, M. Developing and modeling a plane 3 DOF compliant micromanipulator by means of a dedicated MBS code. In Proceedings of the 2011 NSTI Nanotechnology Conference and Expo, Boston, MA, USA, 13–16 June 2011; Volume 2, pp. 659–662.
14. Balucani, M.; Belfiore, N.P.; Crescenzi, R.; Verotti, M. The development of a MEMS/NEMS-based 3 D.O.F. compliant micro robot. *Int. J. Mech. Control* **2011**, *12*, 3–10.
15. Belfiore, N.P.; Balucani, M.; Crescenzi, R.; Verotti, M. Performance analysis of compliant mems parallel robots through pseudo-rigid-body model synthesis. In Proceedings of the ASME 2012 11th Biennial Conference on Engineering Systems Design and Analysis, Nantes, France, 2–4 July 2012; Volume 3, pp. 329–334.
16. Belfiore, N.P.; Emamimeibodi, M.; Verotti, M.; Crescenzi, R.; Balucani, M.; Nenzi, P. Kinetostatic optimization of a MEMS-based compliant 3 DOF plane parallel platform. In Proceedings of the ICCC 2013–IEEE 9th International Conference on Computational Cybernetics, Tihany, Hungary, 8–10 July 2013; pp. 261–266.
17. Sanò, P.; Verotti, M.; Bosetti, P.; Belfiore, N.P. Kinematic Synthesis of a D-Drive MEMS Device with Rigid-Body Replacement Method. *J. Mech. Des. Trans. ASME* **2018**, *140*, 075001. [CrossRef]
18. Verotti, M.; Crescenzi, R.; Balucani, M.; Belfiore, N.P. MEMS-based conjugate surfaces flexure hinge. *J. Mech. Des. Trans. ASME* **2015**, *137*, 012301. [CrossRef]
19. Belfiore, N.P.; Broggiato, G.; Verotti, M.; Balucani, M.; Crescenzi, R.; Bagolini, A.; Bellutti, P.; Boscardin, M. Simulation and construction of a MEMS CSFH based microgripper. *Int. J. Mech. Control* **2015**, *16*, 21–30.
20. Belfiore, N.P.; Broggiato, G.; Verotti, M.; Crescenzi, R.; Balucani, M.; Bagolini, A.; Bellutti, P.; Boscardin, M. Development of a MEMS technology CSFH based microgripper. In Proceedings of the 23rd International Conference on Robotics in Alpe-Adria-Danube Region (RAAD), Smolenice, Slovakia, 3–5 September 2014.
21. Crescenzi, R.; Balucani, M.; Belfiore, N.P. Operational characterization of CSFH MEMS technology based hinges. *J. Micromech. Microeng.* **2018**, *28*. [CrossRef]

22. Di Giamberardino, P.; Bagolini, A.; Bellutti, P.; Rudas, I.; Verotti, M.; Botta, F.; Belfiore, N. New MEMS tweezers for the viscoelastic characterization of soft materials at the microscale. *Micromachines* **2017**, *9*, 15. [CrossRef]

23. Bagolini, A.; Ronchin, S.; Bellutti, P.; Chistè, M.; Verotti, M.; Belfiore, N.P. Fabrication of Novel MEMS Microgrippers by Deep Reactive Ion Etching With Metal Hard Mask. *J. Microelectromech. Syst.* **2017**, *26*, 926–934. [CrossRef]

24. Dochshanov, A.; Verotti, M.; Belfiore, N. A Comprehensive Survey on Microgrippers Design: Operational Strategy. *J. Mech. Des. Trans. ASME* **2017**, *139*, 070801. [CrossRef]

25. Verotti, M.; Dochshanov, A.; Belfiore, N.P. A Comprehensive Survey on Microgrippers Design: Mechanical Structure. *J. Mech. Des. Trans. ASME* **2017**, *139*, 060801. [CrossRef]

26. Verotti, M.; Dochshanov, A.; Belfiore, N.P. Compliance Synthesis of CSFH MEMS-Based Microgrippers. *J. Mech. Des. Trans. ASME* **2017**, *139*. [CrossRef]

27. Potrich, C.; Lunelli, L.; Bagolini, A.; Bellutti, P.; Pederzolli, C.; Verotti, M.; Belfiore, N.P. Innovative silicon microgrippers for biomedical applications: Design, mechanical simulation and evaluation of protein fouling. *Actuators* **2018**, *7*, 12. [CrossRef]

28. Veroli, A.; Buzzin, A.; Crescenzi, R.; Frezza, F.; de Cesare, G.; D'Andrea, V.; Mura, F.; Verotti, M.; Dochshanov, A.; Belfiore, N.P. Development of a NEMS-Technology Based Nano Gripper. In Proceedings of the International Conference on Robotics in Alpe-Adria Danube Region, Torino, Italy, 21–23 June 2017; Springer: Cham, Switzerland, 2017; pp. 601–611.

29. Buzzin, A.; Veroli, A.; de Cesare, G.; Belfiore, N. Nems-Technology based nano gripper for mechanic manipulation in space exploration mission. *Adv. Astronaut. Sci.* **2018**, *163*, 61–67.

30. Nascetti, A.; Caputo, D.; Scipinotti, R.; de Cesare, G. Technologies for autonomous integrated lab-on-chip systems for space missions. *Acta Astronaut.* **2016**, *128*, 401–408. [CrossRef]

31. Hou, M.T.K.; Huang, J.Y.; Jiang, S.S.; Yeh, J.A. In-plane rotary comb-drive actuator for a variable optical attenuator. *J. Micro/Nanolithogr. MEMS MOEMS* **2008**, *7*, 043015.

32. Cho, S.; Chasiotis, I. Elastic properties and representative volume element of polycrystalline silicon for MEMS. *Exp. Mech.* **2007**, *47*, 37–49. [CrossRef]

33. Sharpe, W.N.; Yuan, B.; Vaidyanathan, R.; Edwards, R.L. Measurements of Young's modulus, Poisson's ratio, and tensile strength of polysilicon. In Proceedings of the IEEE The Tenth Annual International Workshop on Micro Electro Mechanical Systems. An Investigation of Micro Structures, Sensors, Actuators, Machines and Robots, Nagoya, Japan, 26–30 January 1997; pp. 424–429.

34. Sharpe, W.N., Jr.; Yuan, B.; Vaidyanathan, R.; Edwards, R.L. New test structures and techniques for measurement of mechanical properties of MEMS materials. In Proceedings of the Micromachining and Microfabrication, Austin, TX, USA, 13 September 1996; Volume 2880, pp. 78–91.

35. Volland, B.; Heerlein, H.; Rangelow, I. Electrostatically driven microgripper. *Microelectr. Eng.* **2002**, *61*, 1015–1023. [CrossRef]

36. Caputo, D.; Ceccarelli, M.; de Cesare, G.; Nascetti, A.; Scipinotti, R. Lab-on-glass system for DNA analysis using thin and thick film technologies. *MRS Online Proc. Libr. Arch.* **2009**, *1191*, 53–58. [CrossRef]

37. De Cesare, G.; Gavesi, M.; Palma, F.; Riccò, B. A novel a-Si: H mechanical stress sensor. *Thin Solid Films* **2003**, *427*, 191–195. [CrossRef]

38. Caputo, D.; de Cesare, G.; Nardini, M.; Nascetti, A.; Scipinotti, R. Monitoring of temperature distribution in a thin film heater by an array of a-Si: H temperature sensors. *IEEE Sens. J.* **2012**, *12*, 1209–1213. [CrossRef]

39. De Cesare, G.; Nascetti, A.; Caputo, D. Amorphous silicon pin structure acting as light and temperature sensor. *Sensors* **2015**, *15*, 12260–12272. [CrossRef] [PubMed]

40. Tucci, M.; Serenelli, L.; Salza, E.; De Iuliis, S.; Geerligs, L.; Caputo, D.; Ceccarelli, M.; de Cesare, G. Back contacted a-Si: H/c-Si heterostructure solar cells. *J. Non-Cryst. Solids* **2008**, *354*, 2386–2391. [CrossRef]

41. Caputo, D.; de Cesare, G. New a-Si: H two-terminal switching device for active display. *J. Non-Cryst. Solids* **1996**, *198*, 1134–1136. [CrossRef]

42. Asquini, R.; Buzzin, A.; Caputo, D.; de Cesare, G. Integrated Evanescent Waveguide Detector for Optical Sensing. *IEEE Trans. Compon. Packag. Manuf. Technol.* **2018**, *8*, 1180–1186. [CrossRef]

43. Gaspar, J.; Chu, V.; Conde, J. Amorphous silicon electrostatic microresonators with high quality factors. *Appl. Phys. Lett.* **2004**, *84*, 622–624. [CrossRef]

44. Veroli, A.; Mura, F.; Balucani, M.; Caminiti, R. Dose influence on the PMMA e-resist for the development of high-aspect ratio and reproducible sub-micrometric structures by electron beam lithography. *AIP Conf. Proc.* **2016**, *1749*, 020010.

actuators

MDPI

Article

An Image Analysis Approach to Microgrippers Displacement Measurement and Testing

Francesco Orsini [1], Federica Vurchio [1], Andrea Scorza [1,*], Rocco Crescenzi [2] and Salvatore Andrea Sciuto [1]

[1] Department of Engineering, Roma Tre University, via della Vasca Navale, 79-00146 Roma, Italy;
    francesco.orsini@uniroma3.it (F.O.); fed.vurchio@stud.uniroma3.it (F.V.);
    salvatore.sciuto@uniroma3.it (S.A.S.)
[2] Department of Information Engineering, Electronic and Telecommunications, Sapienza University of Rome,
    Via Eudossiana, 18-00184 Roma, Italy; crescenzi@diet.uniroma1.it
*   Correspondence: andrea.scorza@uniroma3.it; Tel.: +39-6-5733-3357

Received: 5 September 2018; Accepted: 20 September 2018; Published: 24 September 2018

Abstract: The number of studies on microgrippers has increased consistently in the past decade, among them the numeric simulations and material characterization are quite common, while the metrological issues related to their performance testing are not well investigated yet. To add some contribution in this field, an image analysis-based method for microgrippers displacement measurement and testing is proposed here: images of a microgripper prototype supplied with different voltages are acquired by an optical system (i.e., a 3D optical profilometer) and processed through in-house software. With the aim to assess the quality of the results a systematic approach is proposed for determining and quantifying the main error sources and applied to the uncertainty estimation in angular displacement measurements of the microgripper comb-drives. A preliminary uncertainty evaluation of the in-house software is provided by a Monte Carlo Simulation and its contribution added to that of the other error sources, giving an estimation of the relative uncertainty up to 3.6% at 95% confidence level for voltages from 10 V to 28 V. Moreover, the measurements on the prototype device highlighted a stable behavior in the voltage range from 0 V to 28 V with a maximum rotation of 1.3° at 28 V, which is lower than in previous studies, likely due to differences in system configuration, model, and material. Anyway, the proposed approach is suitable also for different optical systems (i.e., trinocular microscopes).

Keywords: microgripper; displacement; measurement; uncertainty; image analysis

1. Introduction

Microgrippers are MEMS technology-based devices, able to manipulate objects with dimensions of the order of 10^{-6} m. The development and diffusion of such devices is increased considerably thanks to the advent of D-RIE (Deep Reaction Ion Etching) manufacture technology, which led the construction even on a micrometer scale. The microgrippers under study in this article are made up of flexible beams, and the mechanism that allows the movement of these devices is based on a particular and innovative hinge, called a conjugate surface flexure hinge (CSFH), which can be built as a monolithic body and easily integrated into any mechanical structure of a MEMS. These devices are actuated by innovative capacitive rotary comb-drives that can generate a torque when driven by an applied voltage (Figure 1). Based on the theoretical model in [1], to perform an electro-mechanical characterization of these devices, in [2] the behavior of Comb-Drive as studied when a potential difference is applied. In this study, however, some important characteristics such as mass and material effects were neglected and the rotary capacitors were assumed as parallel plates. To validate the theoretical results and obtain a more realistic and accurate model, a numerical simulation based on Finite Element Method (FEM) was also proposed. Moreover, through the Direct Method, it was possible to validate the theoretical behavior, solving the problem of offering a good angle of rotation while minimizing the maximum internal stress. Through a Finite Element Analysis (FEA) simulator, the displacement of the clamps according to the applied potential difference [3] were studied. In addition, in [4], simulations have been conducted on V-Rotation Comb-Drive and V-Force relationships. Thanks to these studies a quadratic behavior of the relationship between the applied voltage and the angular rotation of the Comb-Drive was determined. Despite the great variety of applications involving microgrippers, such as biology [5,6], surgery [7,8] or aerospace [9,10], very few works reported measurement data for the experimental characterization of such devices [2] and the quality of the results is not established through an in-depth uncertainty analysis. To give a contribution to this issue, in this study an experimental approach to measure the microgripper displacements is proposed and applied to verify the theoretical relationship between the applied voltage and the Comb-Drive rotation. To this aim an uncertainty analysis of the quantities measured is also performed: an example of measurements and testing on a microgripper prototype is described here to confirm the theoretical displacement of its comb-drives by means of an optical system (profilometer) and an in-house image analysis software.

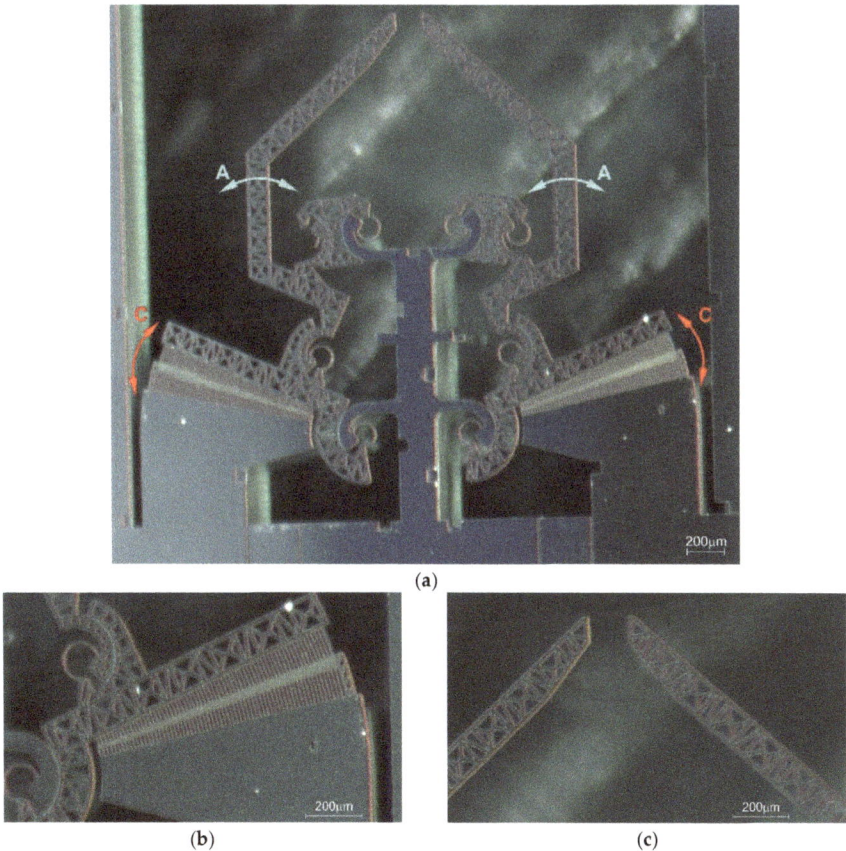

(a)

(b)

(c)

Figure 1. Images of a microgripper captured by a trinocular microscope. (**a**) microgripper, the displacement of the arm A is provided by the rotation of the Comb-Drive C (**b**) Hinges and rotary Comb-Drive (**c**) pliers.

2. Materials and Methods

With the aim to properly characterize and test the microgripper behavior, many kinematic and dynamic output quantities should be measured (e.g., from position and displacement to forces and torques) and related to the input quantities supplied to the device (e.g., voltage, current), therefore a first systematic approach may consider the displacement of one or more of its components taking into account the characteristics of the experimental set up (Table 1).

Table 1. Experimental setup.

Device	Characteristics
Power supply	Keithley 236, Range settable to 1.1/11/110/1100 V with respectively 0.1/1/10/100 mV resolution, accuracy 0.06 V at F.S.
Micropositioners	n.1 MP25L, n.1 MP25R, range X/Y/Z 10/10/10 mm with 5 μm resolution
Probes (supply)	PA-C-1M with tungsten needles
Device Under Test (DUT)	Silicon microgripper (Silicon type P. dopant Boron, orientation <100>, electrical resistivity 0.005–0.03 Ohm×cm), module dimensions 2000 μm × 1500 μm, device thickness 40 μm, insulated layer thickness 3 μm, handle thickness 400 μm, capacitive comb-drives
DUT stage	The wafer containing the DUT is placed on the profilometer working surface fixed by an adhesive tape
Optics: 3D Optical Profilometer	field of view from 7.2 mm × 5.4 mm to 80 nm × 60 nm, maximal lateral resolution 0.6 μm
Digital Image	768 × 580 pixels, 8 bit, 0.6 px/μm
Image Processing Software	In-house software developed in MATLAB (2017a, MathWorks)
Notebook pc	Intel core i7-2670, 6 Gb RAM, Nvidia GeForce GT 520 MX

In particular, with reference to Table 1 the following components have been considered:

a. The microgripper under study (Device Under Test or DUT), made up of flexible beams and hinges, is actuated by capacitive rotary comb-drives that generate a torque when a voltage is applied (Figure 1).
b. A constant voltage source should be provided to evaluate the dependencies of the mechanical output on the voltage. In our study, the power supply device is a Keithley 236 SMU with 0.055% F.S. output voltage accuracy.
c. Two micropositioners embedded with probe arms and tungsten needles have been used (Figure 2) to apply the voltage to the DUT.
d. The microgripper displacements have been measured from images acquired and collected by a Fogale Zoomsurf 3D optical profiling system (Figure 3). The maximum resolution is limited by the diffraction as in conventional optical microscopes to 0.6 μm and the vertical resolution reaches 0.1 nm. The digital image resolution is 0.6 pixel/μm therefore each pixel corresponds to about 1.6 μm. Taking into account that to clearly discriminate 2 point objects in a digital image, 3 pixel are needed in the best case [11,12], the total lateral accuracy into the digital image has been estimated to be ±2 pixel (about ±3.3 μm).
e. The acquired images have been processed by an in-house software developed in MATLAB® and a notebook PC. From the analysis of the image sequences, the movement of the components has been measured by means of a template-matching algorithm. A set of images has been acquired, each one corresponding to a specific voltage setting: to calculate the angular displacement,

the first acquired image (0 V) has been compared with the others (i.e., 2 V, 4 V, ..., 28 V) through the following steps:

- Step 1:

  - The initial coordinates of the Comb-drive ICR (Instant Center of Rotation) are estimated. As shown in Figure 4, four points are manually selected on the edges of the Comb-Drive, and the intersection of the two corresponding lines coincided with the ICR of the Comb-Drive. The second phase involves the manual selection of a particular Region of Interest (ROI) on the image. The Comb-Drive under examination, has been designed in such a way that it has a static part anchored to the structure of the MEMS device and a mobile part. In Figure 4 the lower part is the static one. As shown in Figure 5, a ROI can be selected on the mobile part of device's Comb-Drive to determine the angular displacement (see step 2). Figure 6 shows the superposition of two images in the sequence, one corresponding to no actuation (the neutral configuration) and one corresponding to a 28 V applied voltage.

- Step 2:

  - The coordinates of the following three points are used: the most distant point from the ICR on the fixed Comb-Drive part, the ICR of the Comb-Drive and the center of the ROI. They determined a triangle, where the vertex ICR corresponds to the angular opening of the Comb-Drive.
  - A match is found between the coordinates of the selected ROI on the first image (see step 1) and on all the subsequent images. Through this operation the in-house software can detect the new coordinates of the ROI's center of gravity for each subsequent image and therefore for each applied voltage. On the hypothesis of no deformation due to the movement of the Comb-Drive, the rigid movement of the device keeps unchanged the points on the bottom left and that corresponding to the ICR. Therefore, for each subsequent image in the sequence, the coordinates of the center of the ROI are detected and used to identify a new triangle and consequently a new angular opening of the Comb-Drive for each applied voltage.
  - From the angular aperture of the Comb-Drive at each voltage, the corresponding angular displacement is obtained.

The above steps can be easily adapted to measure the movement of other elements of the microgripper and are suitable on digital images produced also by systems different from that used in this work. On the other hand, it is important to consider the errors due to the variability of the parameters related to the manual selection of the operator, together with the uncertainty of the code itself. For this reason, an uncertainty analysis has been made considering the above issues by means of Monte Carlo Simulation (MCS).

Figure 2. Micro-positioner.

Figure 3. Optical profiling system.

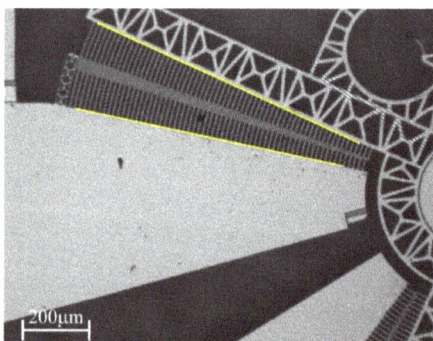

Figure 4. Microgripper under study (DUT): image of the Comb-Drive provided by the 3D optical profiling system. From the intersection of the yellow lines the ICR is determined.

Figure 5. Image of the DUT provided by the 3D optical profiling system. ROI identification on the mobile arm of the Comb-Drive.

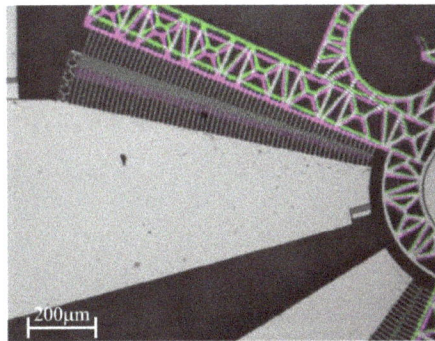

Figure 6. Superposition of two images of the DUT Comb-Drive, corresponding to 0 V and 28 V applied voltage, respectively.

First of all, an uncertainty analysis of the errors due to the experimental setup has been done, therefore the main uncertainty sources have been identified and evaluated, as shown in Table 2 where for each source a probability density function PDF together to the uncertainty type, mean m and standard deviation σ is determined.

- Power supply uncertainty has been evaluated trough the data sheet and a Gaussian distribution has been assumed, therefore the corresponding standard deviation σ is evaluated from the declared uncertainty u as $\sigma = u/2$.
- DUT stage: orientation and vibrations of the DUT stage are related to a change in image focusing during measurements. The corresponding uncertainty has been assumed negligible as the setup is mounted on an optical table with a pneumatic suspension system and at each step a correct image focusing is maintained.
- Optical system and digital image: as described before the uncertainty due to the use of the optical profilometer has been evaluated considering its maximal lateral resolution and the digital resolution of the acquired image, i.e., 0.6 μm and 1.6 μm/px, respectively. As shown in Figures 7–10, two point objects in a digital image are resolved if they are separated by 1 or more pixel depending if they are centered or not on a picture element. In the best case (i.e., for high contrast and Signal to Noise Ratio) the minimum distance needed between them should be about 3 pixels [11,12], therefore in this study a uniform uncertainty distribution with a 2 pixels semi-amplitude is assumed.

- Software Processing: two uncertainty contributions can be determined: the former is related to the step 1 (manual selection), while the other is due to the template-matching algorithm. As in previously works [14–17] the total uncertainty has been evaluated by an MCS.

Table 2. Uncertainty sources.

Uncertainty Source	Probability Density Function (PDF) [1]	Type [2]	m	σ	Unit
Power Supply: Voltage accuracy	N (m, σ)	B		0.03	V
DUT Stage: Plate planarity and vibration, focusing plane variation	-	-	-	-	-
Optical System: Profilometer Maximal lateral resolution due to diffraction	U (m, σ)	B		0.4	μm
Digital Image: digital conversion	U (m, σ)	B		1.9	μm
Calibration: Included in the nominal uncertainty of the Optical System	-	-	-	-	-
Image Processing Software: uncertainty in point identification uncertainty in ROI position (template-matching) uncertainty in ROI size (template-matching)	U (m, σ)	A		±2 ±2 ±2	pixel pixel pixel
uncertainty in the template-matching algorithm	A Monte Carlo Simulation is used.				

[1] N(m, σ) is a Gaussian PDF with mean m and standard deviation σ; U(m, σ) is a uniform PDF with mean m and standard deviation σ; [2] Type A and type B uncertainty as in [13]

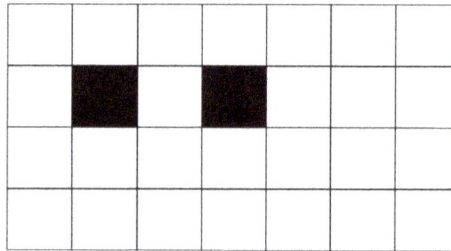

Figure 7. Two pixel-centered point elements separated by 2 pixels.

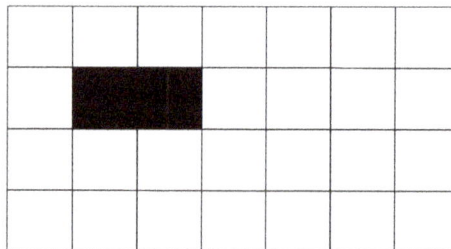

Figure 8. Two point elements in 2 near pixels.

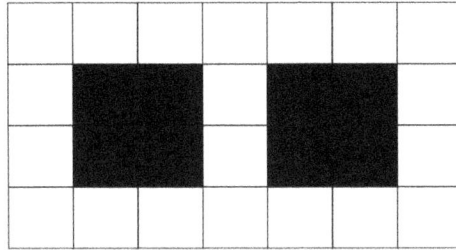

Figure 9. Two point elements not pixel-centered separated by 3 pixels.

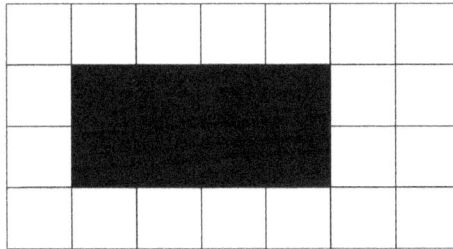

Figure 10. Two point elements not pixel-centered separated by 2 pixels.

To evaluate the accuracy of image processing software, a simulation was implemented in MATLAB® in accordance with the MCS Method. In this procedure the uncertainty sources were introduced by the variability of the parameters related to the manual selection of the operator. In particular:

- Uncertainty due to point selection: with the aim to determine the ICR, the operator manually selects some points on the image. This operation is performed only once at the beginning of the measurement process, and even if the coordinates of these points are confined to a limited region of space (the mobile edge of the Comb-Drive), there is a variability introduced by the nature of the manual process. Under the same conditions (same image selected and same operator), a set of 10 consecutive tests were carried out and a dispersion of ±2 pixels was estimated. A uniform distribution U (m, σ) was also assumed with semi-amplitude of 2 pixels and mean m from the average of the 10 tests.

- Uncertainty related to ROI position in the template-matching algorithm: the operator arbitrarily chooses the ROI position in the image. Although the ROI lies in a limited region of the image (i.e., the mobile part of the Comb-Drive), an uncertainty is introduced by the variability of the manual selection. Also, in this case, a set of 10 consecutive tests were carried out and a dispersion of ±2 pixels was determined: a uniform distribution U (m, σ) was assumed with a 2-pixel semi-amplitude and mean m from the average of the 10 tests.

- Uncertainty related to ROI size in the template-matching algorithm: in the same way, the ROI size selected by the operator may vary. Under the same conditions (same image selected and same operator), also in this case a set of 10 consecutive tests were carried out and a ±2 pixels dispersion was estimated. A uniform distribution U (m, σ) was also assumed with mean value equal to the average of the 10 trials and a 2-pixel semi-amplitude.

- Uncertainty of template-matching algorithm: also, the template-matching gives a contribution to uncertainty as described in [13], nevertheless it is included in the MCS results, as the final data dispersion is affected by this source of error too.

Once the distributions related to the variability of manual selection processes have been established, the process of calculating the angular displacement of the Comb-Drive is performed 10,000 times (i.e., 10,000 iterations in the MCS), for an applied potential difference of 28 V. This procedure is carried out to verify the distribution of the angular displacements when the above parameters change.

The total uncertainty u_c of the displacement measurement is evaluated as suggested in [13] with the following expression:

$$u_c = \sqrt{u_A^2 + u_B^2} \tag{1}$$

where u_A and u_B are the type A and type B overall uncertainty respectively. In particular, u_A is the uncertainty contribution evaluable with a statistic analysis of the measurements dispersion obtained with the system and u_B is the overall uncertainty from the main error sources related to the experimental setup. To also evaluate the uncertainty contribution of the power supply the relation f between the voltage V and the angular displacement α has been considered. The u_B can be written as:

$$u_B = \delta\alpha = \sqrt{(\delta\alpha_t)^2 + \left(\frac{\partial f}{\partial V}\delta V\right)^2} \qquad \frac{\delta\alpha_t}{\alpha_t} = \sqrt{\left(\frac{\delta\alpha_p}{\alpha}\right)^2 + \left(\frac{\delta\alpha_s}{\alpha}\right)^2} \tag{2}$$

where α_p are the angle measurement uncertainty due to the profilometer system (optical and A/D conversion), α_s is the angle measurement uncertainty due to the implemented software system, δV the uncertainty on the supply voltage and $\partial f/\partial V$ is the derived function that described the correlation between the supply voltages and the Comb-Drive rotations [13]. This last uncertainty contribution has been evaluated through an MCS. To evaluate the α_p contribution, the angle has been measured by means of the triangular properties as shown in Figure 11 and (3).

$$\alpha_p \simeq tg\alpha_p = \frac{a}{b} \tag{3}$$

From Figure 11 the values a and b are obtained from the Δx and Δy lengths, following the expressions in (4):

$$a = \sqrt{\Delta x_a^2 + \Delta y_a^2}, \ b = \sqrt{\Delta x_b^2 + \Delta y_b^2} \tag{4}$$

Therefore, the uncertainty due to the profilometer error in Table 2 can be obtained as:

$$\frac{\delta\alpha_p}{\alpha_p} = \sqrt{\left(\frac{\delta a}{a}\right)^2 + \left(\frac{\delta b}{b}\right)^2} \tag{5}$$

where:

$$\delta a = \sqrt{\left(\frac{\partial a}{\partial x}\delta x\right)^2 + \left(\frac{\partial a}{\partial y}\delta y\right)^2} = \sqrt{\left(\frac{\Delta x_a}{\sqrt{\Delta x_a^2 + \Delta y_a^2}}\delta x\right)^2 + \left(\frac{\Delta y_a}{\sqrt{\Delta x_a^2 + \Delta y_a^2}}\delta y\right)^2} \tag{6a}$$

$$\delta b = \sqrt{\left(\frac{\partial b}{\partial x}\delta x\right)^2 + \left(\frac{\partial b}{\partial y}\delta y\right)^2} = \sqrt{\left(\frac{\Delta x_b}{\sqrt{\Delta x_b^2 + \Delta y_b^2}}\delta x\right)^2 + \left(\frac{\Delta y_b}{\sqrt{\Delta x_b^2 + \Delta y_b^2}}\delta y\right)^2} \tag{6b}$$

The quantities a and b in (6) depend on the image resolution and size. For this study, the lengths of the Comb-Drive triangle in the image are considered for the maximum rotation, i.e., $\Delta x_a = 25$ pixel, $\Delta y_a = 150$ pixel, $\Delta x_b = 680$ pixel and $\Delta y_b = 150$ pixel, therefore $a = 152$ pixel, and $b = 700$ pixel.

From Table 2 and the maximum measured angle is 1.3° and the relative uncertainty is up to 3.6% for voltages higher than 10 V.

In a second step of our study some tests have been conducted to evaluate how good the microgripper behavior fits the theoretical model [2]. With this aim it has been supplied with different voltages, i.e., from 0 V to 28 V at 2 V steps, forward and backward 6 times.

In Figure 6 two superimposed images of the same Comb-Drive at different voltages are shown: the in-house software can evaluate the angle difference between them.

For each voltage value the standard deviation and the average of the angular displacement of the Comb-Drive have been measured as above. Finally, all the measurement data have been processed and interpolated to provide a curve fitting of the actual motion of the Comb-Drive depending on the voltage supply.

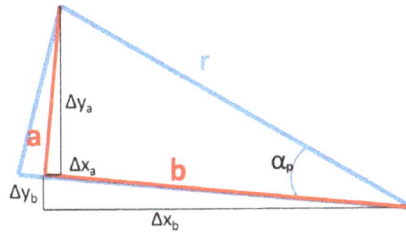

Figure 11. Angle measurement for the uncertainty evaluation related to the profilometer image.

3. Results

From the MCS the image processing software accuracy has been estimated: as shown in Figure 12 the distribution of results followed a Gaussian curve with $\pm 0.02°$ interval at a 95% confidence level (percentile values for 2.5% and 97.5% are 1.29° and 1.32° respectively). This value has been used to evaluate the total measuring system uncertainty following (1).

The uncertainty analysis above has been applied to angular measurements on the microgripper at different voltages (Figure 13, Table 3). Measurement data showed that the interpolating curve (voltage vs angular displacement) was similar to the numerical results and to measurements reported in other works on similar devices [2,16], but the angles measured here were usually lower for the same voltages. This different behavior is likely due to the different microgripper used (a different model, configuration, and material). Indeed, in the studies above mentioned [2,16], different microgripper configurations and models provided different angular behaviors, as shown in Table 3.

In Figure 13 the measurement results proposed here seem to be more similar to [16] than [2]. The different materials and configuration may be related to different elastic resistances in the hinges translating in different angular displacements. Anyway, a best fitting curve with a $R^2 = 0.999$ demonstrated a coherent behavior along all the voltages: this is likely due also to the systematic approach and to the in-house software applied in this study to perform the measurements with repeatability and low-operator-dependence.

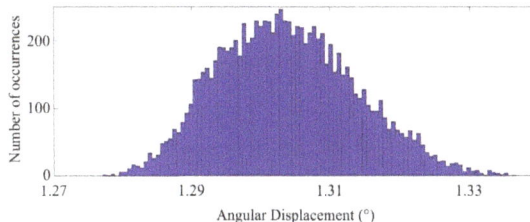

Figure 12. Example of results of the Monte Carlo Simulation for software uncertainty analysis (10,000 cycles).

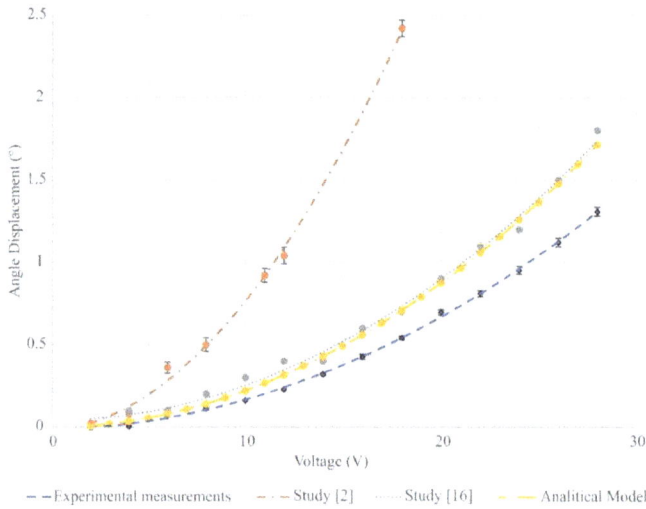

Figure 13. Curve and points obtained for measurements made with the proposed system (blue dashed line) compared with that from the analytical model (yellow dashed line) and the experimental data (orange dot-dashed line) in [2,16] (dotted line).

Table 3. Results on angular displacement measurements (Figure 13).

Supply Voltage (V)	Angle Rotation Mean Val. (°)	Standard Deviation (°)	Angle [2] (°)	Angle [16] (°)
2	0.005	0.005	0.02	0
4	0.011	0.008	0.08	0.1
6	0.065	0.004	0.24	0.1
8	0.11	0.011	0.5	0.2
10	0.1603	0.004	0.8	0.3
12	0.227	0.006	1.08	0.4
14	0.3190	0.007	1.65	0.4
16	0.427	0.015	2.08	0.6
18	0.5425	0.011	-	0.7
20	0.696	0.016	-	0.9
22	0.810	0.018	-	1.1
24	0.952	0.024	-	1.2
26	1.122	0.027	-	1.5
28	1.308	0.027	-	1.8

4. Conclusions

With the aim to characterize and test the behavior of a microgripper a systematic approach to measurements is proposed and applied. An experimental setup has been developed and the main sources of error have been evaluated to estimate the measurement uncertainty in angular displacements of the microgripper Comb-Drive depending on the applied voltage.

The results showed a stable behavior of the Comb-Drive with angular displacement lower than in other works, that was likely due to the different MEMS configuration and model, as the different materials and configuration could imply a different elastic resistance in the hinges evolving in different angular displacement.

The software here implemented can work on different components of the gripper (clamp instead Comb-Drive). Moreover, the method proposed can be easily applied to images provided by systems that provide bidimensional images of the microgripper in operational conditions, e.g., it can work also with images from a trinocular microscope equipped in a probe station. In this case, an in-depth analysis of the probe station characteristics and its calibration should be done. Although the overall measurement uncertainty may be higher than in the case described here, it must be pointed out that the optical microscope is a measurement system less complex and expensive than the 3D optical profilometer, therefore it may be a good trade-off between costs and results accuracy. Future works will be focused on this kind of testing system and a study on the clamp motion behavior in comparison with the Comb-Drive rotation is going to be conducted to further contribute to a more complete approach to microgripper experimental characterization and testing.

Author Contributions: Conceptualization, A.S.; Methodology, F.O., F.V. and A.S.; Software, F.O. and F.V.; Validation, F.O., F.V. and A.S.; Formal Analysis, F.O.; Investigation, F.V. and F.O.; Resources, S.A.S. and R.C.; Data Curation, F.V.; Writing-Original Draft Preparation, F.O., F.V. and A.S.; Writing-Review & Editing, A.S., F.O., F.V. and S.A.S.; Visualization, F.O. and F.V.; Supervision, A.S. and S.A.S.; Project Administration, A.S. and S.A.S.; Funding Acquisition, S.A.S.

Conflicts of Interest: The authors declare no conflict of interest.

References

1. Tu, C.C.; Fanchiang, K.; Liu, C.H. 1 × N rotary vertical micromirror for optical switching applications. In Proceedings of the MOEMS-MEMS Micro and Nanofabrication, San Jose, CA, USA, 22 January 2005.
2. Crescenzi, R.; Balucani, M.; Belfiore, N.P. Operational characterization of CSFH MEMS-Technology based hinges. *J. Micromech. Microeng.* **2018**, *28*, 055012. [CrossRef]
3. Cecchi, R.; Verotti, M.; Capata, R.; Dochshanov, A.; Broggiato, G.B.; Crescenzi, R.; Balucani, M.; Natali, S.; Razzano, G.; Lucchese, F.; et al. Development of Micro-Grippers for Tissue and Cell Manipulation with Direct Morphological Comparison. *Micromachines* **2015**, *6*, 1710–1728. [CrossRef]
4. Veroli, A.; Buzzin, A.; Crescenzi, R.; Frezza, F.; de Cesare, G.; D'Andrea, V.; Mura, F.; Verotti, M.; Dochshanov, A.; Belfiore, N.P. Development of a NEMS-Technology Based Nano Gripper. In Proceedings of the International Conference on Robotics in Alpe-Adria Danube Region, Torino, Italy, 21–23 June 2017.
5. Panescu, D. Mems in medicine and biology. *IEEE Eng. Med. Biol. Mag.* **2006**, *25*, 19–28. [CrossRef] [PubMed]
6. Kim, K.; Liu, X.; Zhang, Y.; Sun, Y. Nanonewton force-controlled manipulation of biological cells using a monolithic mems microgripper with two-axis force feedback. *J. Micromech. Microeng.* **2008**, *18*, 055031. [CrossRef]
7. Gosline, A.; Vasilyev, N.; Butler, E.; Folk, C.; Cohen, A.; Chen, R.; Lang, N.; Del Nido, P.; Dupont, P. Percutaneous intracardiac beating-heart surgery using metal mems tissue approximation tools. *Int. J. Robot. Res.* **2012**, *31*, 1081–1093. [CrossRef] [PubMed]
8. Benfield, D.; Yue, S.; Lou, E.; Moussa, W. Design and calibration of a six-axis mems sensor array for use in scoliosis correction surgery. *J. Micromech. Microeng.* **2014**, *24*, 085008. [CrossRef]
9. Leclerc, J. MEMS for aerospace navigation. *IEEE Aerosp. Electron. Syst. Mag.* **2007**, *22*, 31–36. [CrossRef]
10. Ibrahim, D. Using MEMS accelerometers in aerospace and defence electronics. *Electron. World* **2012**, *118*, 16–21.
11. Pawley, J.B. Points, Pixels and Gray Levels: Digitizing Image Data. In *Handbook of Biological Confocal Microscopy: Third Edition*; Springer Publishing: New York, NY, USA, 2006; pp. 59–79.
12. Waters, J.C. Accuracy and precision in quantitative fluorescence microscopy. *J. Cell Biol.* **2009**, *185*, 1135–1148. [CrossRef] [PubMed]
13. ISO/IEC Guide 98-3: 2008. Available online: https://www.iso.org/standard/50461.html (accessed on 21 September 2018).

14. Orsini, F.; Scena, S.; D'Anna, C.; Scorza, A.; Schinaia, L.; Sciuto, S.A. Uncertainty Evaluation of a Method for the Functional Reach Test Evaluation by means of Monte-Carlo Simulation. In Proceedings of the 22nd IMEKO TC4 International Symposium & 20th International Workshop on ADC Modelling and testing supporting world development through electrical & electronic measurements, Iasi, Romania, 14–15 September 2017.

15. Orsini, F.; Scorza, A.; Rossi, A.; Botta, F.; Sciuto, S.A.; Di Giminiani, R. A preliminary uncertainty analysis of acceleration and displacement measurements on a novel WBV platform for biologic response studies. In Proceedings of the 2016 IEEE International Symposium on Medical Measurements and Applications (MeMeA), Benevento, Italy, 15–18 May 2016.

16. Bagolini, A.; Ronchin, S.; Bellutti, P.; Chistè, M.; Verotti, M.; Belfiore, N.P. Fabrication of Novel MEMS Microgrippers by Deep Reactive Ion Etching with Metal Hard Mask. *J. Microelectromech. Syst.* **2017**, *26*, 926–934. [CrossRef]

17. Scorza, A.; Pietrobon, D.; Orsini, F.; Sciuto, S.A. A preliminary study on a novel phantom based method for performance evaluation of clinical colour doppler systems. In Proceedings of the 22nd IMEKO TC4 International Symposium & 20th International Workshop on ADC Modelling and Testing Supporting World Development Through Electrical & Electronic Measurements, Iasi, Romania, 14–15 September 2017.

actuators

MDPI

Article

Scalloping and Stress Concentration in DRIE-Manufactured Comb-Drives

Silvia Bertini [1], Matteo Verotti [2], Alvise Bagolini [3], Pierliugi Bellutti [3], Giuseppe Ruta [1,4] and Nicola Pio Belfiore [5,*,†]

[1] Sapienza University of Rome, Dept. Structural and Geotechnical Engineering, 00184 Rome, Italy; silvia.bertini@mail.com (S.B.); giuseppe.ruta@uniroma1.it (G.R.)
[2] Università degli studi Niccolò Cusano, 00166 Rome, Italy; matteo.verotti@unicusano.it
[3] MNF-CMM Kessler Foundation for Research, 38123 Trento, Italy; bagolini@fbk.eu (A.B.); bellutti@fbk.eu (P.B.)
[4] Gruppo Nazionale di Fisica Matematica, 00185 Rome, Italy
[5] Department of Engineering, Università degli studi di Roma Tre, 00154 Rome, Italy
* Correspondence: nicolapio.belfiore@uniroma3.it; Tel.: +36-06-5733-3316
† Current address: Dipartimento di Ingengeria, via della vasca navale 79, 00146 Rome, Italy.

Received: 23 July 2018; Accepted: 4 September 2018; Published: 5 September 2018

Abstract: In the last decades, microelectromechanical systems have been increasing their number of degrees of freedom and their structural complexity. Hence, most recently designed MEMSs have required higher mobility than in the past and higher structural strength and stability. In some applications, device thickness increased up to the order of tens (or hundred) of microns, which nowadays can be easily obtained by means of DRIE Bosch process. Unfortunately, scalloping introduces stress concentration regions in some parts of the structure. Stress concentration is a dangerous source of strength loss for the whole structure and for comb-drives actuators which may suffer from side pull-in. This paper presents an analytical approach to characterize stress concentrations in DRIE micro-machined MEMS. The method is based on the linear elasticity equations, the de Saint-Venant Principle, and the boundary value problem for the case of a torsional state of the beam. The results obtained by means of this theoretical method are then compared with those obtained by using two other methods: one based on finite difference discretization of the equations, and one based on finite element analysis (FEA). Finally, the new theoretical approach yields results which are in accordance with the known value of the stress concentration factor for asymptotically null radius notches.

Keywords: scalloping; stress concentration; compliant mechanisms; DRIE

1. Introduction

Interest in microelectromechanical systems devices has steadily increased in the last years. Micro-scale technologies have proven to be very effective, playing a prominent role in a wide variety of applications from different fields, such as drug delivery [1–5], aerospace [6,7], medical diagnosis [8,9], surgical applications [10–12], and cell manipulation [13–15].

Since 2013 [16,17], a new class of MEMS, equipped with Conjugate Surface Flexure Hinges (CSFH) has been developed [18,19] and fabricated [20]. A sample of CSFH is depicted in Figure 1. These systems consist essentially of micro *compliant mechanisms* where the flexure hinges are manufactured as CSFHs. In such kind of flexure, the thickest *rigid* elements are linked by a certain number of CSFHs, where each CSFH consists of a curved beam together with a portion of a conjugate-profile pair, as illustrated in Figure 1. According to the CSFH design, the center of each arc is coincident with the center of the elastic weights of the curved beam [21].

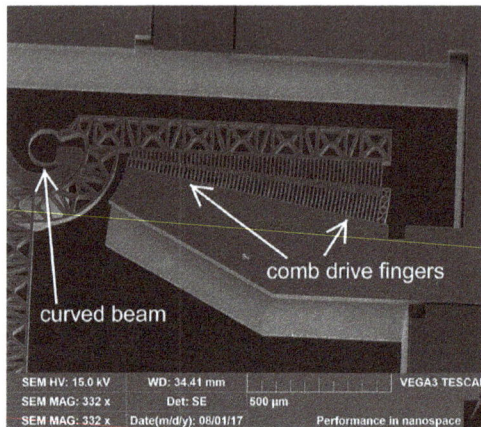

Figure 1. Curved beam adopted in CSHF hinges and comb-drive fingers.

The availability of CSFH hinges gave rise to the design and fabrication of new devices, such as microrobots [16,17,22–24] micromechanisms [17,25], microgrippers [18,26–28], and even a microtribometer [29].

In-plane MEMSs perform properly only if their flexible elements offer high compliance selectivity around the perpendicular axis. Hence, high aspect ratio structures are desirable. While surface micro-machining offers, typically, thickness from 2 to 5 µm, the Deep-Reactive Ion Etching (DRIE) process is used to obtain thickness of 50 µm and much more.

Since 1993 [30], Bosch DRIE process has introduced an effective method to provide, simultaneously, high mask selectively and very high anisotropy of the etched structure. In 1999, the scalloping observed on vertical walls during time multiplexed deep etching (TMDE), the roughness of horizontal surfaces exposed to the glow discharge, and the radius at the bottom of etched features were highlighted [31]. Few years later, it was demonstrated that important features of the samples, such as achievable anisotropy, etching uniformity, fillet radii, and, mainly, surface roughness, are strongly dependent on chamber pressure, applied coil and electrode power, and reactant gases flow rate [32]. This experimental observation lays the groundwork for the present investigation. In fact, scalloped sidewall roughness affects the performance of DRIE-manufactured comb-drives from both an *electrical* and a *mechanical* point of view. An example of DRIE-manufactured rotary comb-drive is depicted in Figure 1.

Considering the electrical aspect, scalloping determines, along the thickness of the device layer, gap variations between the facing surfaces of the fixed and movable fingers. Therefore, it affects the force or the torque exerted by the linear or rotary comb-drive, respectively. Furthermore, charge concentrations occur in sharp edges [33], determining variations of the electric field along the thickness direction.

From a mechanical perspective, it was observed that [31,32] post etch behavior for specimens with high surface roughness always indicated low fracture strength, whereas for specimens with better surface quality there was a wider distribution in sample strength. In addition, recent experimental tests [34] showed that some teeth are subject to lateral pull-in, with possible large deflection and instability.

The present investigation focuses on the mechanical aspect.

As discussed in the next paragraph, one source of strength decay is the stress concentration factor, which holds in those zones where surface discontinuities appear. In high aspect ratio DRIE Bosch process, problems and challenges regarding scalloping attenuation have been reported in various investigations [35–40], although the problem presented several difficulties.

More recently, it has been shown that the sidewall roughness varies with the depth and depends on the trench width. In addition, the surface upper region exhibits a scalloping morphology, while the rougher lower region shows a curtaining morphology [41]. Although cryogenic-DRIE (see for example Refs. [42–45]) has recently been employed to fabricate trenches with highly vertical sidewalls and obtained relatively smooth surfaces, Bosch DRIE remains of great interest for the construction of many devices.

In 2012, a study [46] dedicated to Through-Silicon Via (TSV) technology showed that an etch method not based on Bosch process can get quite smooth sidewalls with no scalloping, together with a good control of the etched profile.

Since scalloping is not acceptable in specific applications, a dedicated effort has been provided to control and reduce it, mainly based on the optimization of DRIE process parameters [47–50], or, alternatively, removing it after fabrication [38].

Unfortunately, indentations and notches on the surface of a DRIE manufactured device give rise to high stress concentration that leads the samples to break at an unexpectedly low load.

The present work intends to contribute in filling this gap and to investigate about the shear stress distribution in the neighborhood of each single scalloped discontinuity on the walls device.

Hence, to develop a theoretical model, a uniform cantilever beam with circular cross section has been considered. To model the scalloping effect, a circular region, or *notch*, has been subtracted from the beam cross section. The center of the notch lies on the cross section profile.

By applying the linear elasticity theory to the beam with notched cross section, the shear stress distribution can be obtained *analytically* for the case of pure torsion. The idea of studying the beam torsional state can be justified by considering that stress concentration is *actually* due to tangential tensions. In addition, the shearing force is quite likely eccentric with respect to the center of shear of the cross-section. This implies a twisting moment acting on the beam, which is dangerous because of the very low torsion rigidity of thin-walled open profiles. Furthermore, the equations are integrated also by using a numerical approach based on finite discretization of the section and finite differences. Finally, the results are compared to those obtained with finite element analysis (FEA).

2. Motivation of This Work

The Bosh DRIE process is performed by means of pulsed regimes in two steps: in the first one, plasma etches the wafer along mainly the vertical direction, while in the second phase a passivation layer (Polytetrafluoroethylene) is chemically deposited on the silicon surface thanks to the action of a source gas (usually Octafluorocyclobutane). The two steps are iterated in such a way that, after each surface passivation phase, an etching phase follows, during which ions attack selectively the passivation layer at the bottom of the trench but not along the side walls. This iterative process permits very deep vertical etching [30]. However, the etching phase affects also the lateral walls and so the side is not perfectly planar, but it is rather undulating, as pictorially represented in Figure 2. The first image, reported in Figure 2a, represents a simplified scheme of a portion of a SOI wafer, after four cycles (note that the stop layer, represented in gray color, is reached after one more cycle only). The central cantilever beam appears after the wet etching phase, as illustrated in Figure 2b,c, during which the oxide is partially removed.

As known, any geometric irregularity on the cross section of a beam is responsible for stress concentration, which decreases its actual strength. The capability of micro-electromechanical systems (MEMS) to move is a direct consequence of their embedded flexures. According to the adopted process, a SOI (Silicon on Insulator) is used to obtain a suspended so-called *device layer* that can move along *in-plane* directions. The geometry of the device is designed and then transferred to the device layer by means of a mask. In the device layer, there are some *pseudo-rigid* parts that achieve motion capability because of the presence of flexible micro-beams. It is therefore clear that the strength of the micro-beams affects performance and reliability of the whole microsystem.

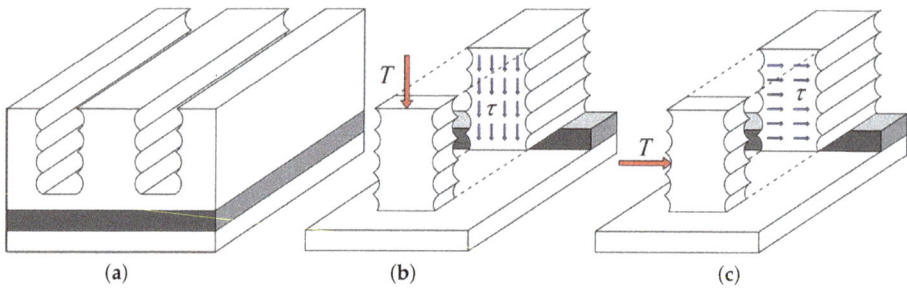

Figure 2. A simplified representation of the SOI wafer after four cycles of the D-RIE process (**a**), and of a cantilever beam after oxide wet etching, subject to out-of-plane (**b**) and in-plane (**c**) shear tension.

Given the large variety of applications, micro-beams in MEMS are used for many different purposes. For example, they can be used either to create the fingers (linear or curved) of a comb-drive actuators or as a structural element (once again linear of curved) of a flexure. Therefore, these micro-beams are generally subject to local or distributed external forces that induce shear tensions on the beam cross-section. These forces may act along either an *out-of-plane* or an *in-plane* direction. Out-of-plane tensions are represented in Figure 2b, while in-plane tensions are depicted in Figure 2c. In the case of comb-drive actuators, fingers are subject to the *lateral* pull-in effect, and therefore they are inflected around an axis that is orthogonal to the pane of motion (in-plane inflection). On the other hand, surface forces, such as electrostatic forces, may act perpendicularly to the plane of motion, giving rise to out-of-plane inflection.

Furthermore, since the gaps among fingers can be reduced to as small as one micron, the possible inflection is responsible for stitching, which makes the MEMS practically impossible to use any further. The mechanics of adhesion this context has been extensively discussed in a recent review [51], where some fundamental parameters for microscale adhesion in MEMS are discussed in detail.

Figure 3 shows scalloping over nine fingers of a circular comb-drive that the authors use for MEMS actuation. Although comb-drives fingers do not present direct contacts, mechanical impacts, vibrations or electrostatic field forces can still play a certain action on the fingers.

Figure 3. Comb-drive finger region.

However, a much greater stress state is generated on the curved beam which sustains the moving parts. For example, in the case of CSFH hinges [34], the curved beam is bent around an axis which is orthogonal to the plane of motion.

Figure 4 shows how scalloping produces roughness on one finger surface.

Figure 4. Scalloping details in one finger.

3. Theoretical Background

We refer to a prismatic beam as a region bounded by an exterior cylindrical surface and two terminal plane sections perpendicular to the generators [52]. The beam axis has a mild curvature compared to its thickness, so the theory for straight beams can be extended to this case of study. According to Ref. [53], this assumption is true when the ratio of the curvature radius to thickness is greater than 5.

The generic cross-section is depicted in Figure 5. Axes x and y, not necessarily principal of inertia, intersect each other in the centroid S and form an orthogonal Cartesian frame with z lying along the direction of the beam axis.

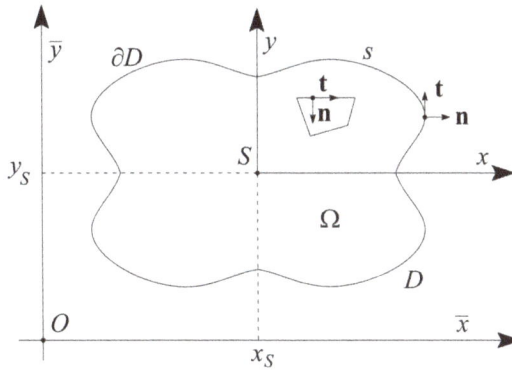

Figure 5. Cross-section with nomenclature.

The present formulation applies with no restriction to cross-sections of arbitrary shape and also to hollow beams. However, the case of simply-connected domain is analyzed here.

In the case of two axes of symmetry, the section centroid coincides with the center of shear, and extension, bending , and torsion are uncoupled if forces are applied to S [54–58].

The cross-section is a domain D of area Ω, bounded by the exterior directrix s whose direction is uniquely defined asking the associated outward normal $\mathbf{n}$ and tangent $\mathbf{t}$ unit vectors to form a right-handed system with the unit vector $\hat{e}_z$.

In the following sections, shear stresses is determined from the theory of linear elasticity. Once the stress and strain tensors are defined, the semi-inverse method of Saint-Venant [59–61] is applied by setting [52,54,55,57,58]

$$\sigma_x = \sigma_y = \tau_{xy} \equiv 0,$$

the equilibrium equations, neglecting body forces, read

$$\frac{\partial \tau_{xz}}{\partial z} = 0, \tag{1}$$

$$\frac{\partial \tau_{yz}}{\partial z} = 0, \tag{2}$$

$$\frac{\partial \sigma_z}{\partial z} + \frac{\partial \tau_{xz}}{\partial x} + \frac{\partial \tau_{yz}}{\partial y} = 0. \tag{3}$$

Assuming the exterior surface to be free of stress corresponds to setting

$$\tau_{xz} dy - \tau_{yz} dx = 0 \tag{4}$$

on ∂D, with $dy = n_x ds$ and $dx = -n_y ds$.

The hypotheses of isotropy and linear elastic behavior of the material lead to the stress-strain relations:

$$\sigma_z = E\epsilon_z = E\frac{\partial w}{\partial z}, \tag{5}$$

$$\tau_{xz} = G\gamma_{xz} = G\left(\frac{\partial u}{\partial z} + \frac{\partial w}{\partial x}\right), \tag{6}$$

$$\tau_{yz} = G\gamma_{yz} = G\left(\frac{\partial v}{\partial z} + \frac{\partial w}{\partial y}\right), \tag{7}$$

where $\mathbf{u} = [u, v, w]^T$ is the small displacement field, and E and G are Young's and shear modulus, respectively.

The stress resultants on the cross-section are the normal and shear forces T_z, T_x, T_y and the twisting and bending couples M_z, M_x and M_y:

$$
\begin{aligned}
T_z &= \int_D \sigma_z \, dxdy, & M_x &= \int_D y\sigma_z \, dxdy, \\
T_x &= \int_D \tau_{xz} \, dxdy, & M_y &= -\int_D x\sigma_z \, dxdy, \\
T_y &= \int_D \tau_{yz} \, dxdy, & M_z &= \int_D (x\tau_{yz} - y\tau_{xz}) \, dxdy.
\end{aligned}
\tag{8}
$$

4. Shear Stresses in Torsion

As mentioned in Section 1, we focus on the shear stress concentration due to a scallop-like geometry in a beam subject to pure torsion. To compute the shear stresses, the solution of Poisson's problems, with Neumann or Dirichlet boundary conditions, is required. Such problems imply the use of Laplace operator in two dimensions [52,54–58,62].

Pure torsion is defined as the partial solution of the problem of shear stresses where $\sigma_z \equiv 0$ (implying the absence of shear forces and bending moments) [52]. The shear stresses $\tau = [\tau_{xz}, \tau_{yz}]^T$ satisfy the simplified equilibrium equation:

$$\text{div } \tau = \frac{\partial \tau_{xz}}{\partial x} + \frac{\partial \tau_{yz}}{\partial y} = 0. \tag{9}$$

When a twisting couple M_z is applied to the free end of a clamped beam, the flatness of the section is preserved only if the beam has circular cross-sections. In all other cases, the longitudinal displacement w is not linear in the variables x and y, causing a warping of the section.

Assuming the torsion angle of the cross-section $\omega_o = \omega_o(z)$ to be small, the displacement field **u** in pure torsion is devoid of rigid contributions,

$$u = -\omega_o(z)y, \quad v = \omega_o(z)x, \quad w = \theta g(x,y), \tag{10}$$

where the constant parameter $\theta = \frac{d\omega_o}{dz}$ is the twist of the centroidal fiber and $g(x,y)$ is the so-called *warping function*.

Generally, the function $g(x,y)$ is not linear in x and y and causes the loss of flatness of the section. It is defined only up to an additive constant and can be made unique by the constraint

$$\int_D g \, dx dy = 0.$$

Deriving shear strains as shown in Equations (6) and (7) allows us to write shear stresses as:

$$\tau_{xz} = G\theta\left(\frac{\partial g}{\partial x} - y\right), \quad \tau_{yz} = G\theta\left(\frac{\partial g}{\partial y} + x\right). \tag{11}$$

By substituting Equation (11) into the equilibrium condition, Equation (9), leads to

$$G\theta\left(\frac{\partial^2 g}{\partial^2 x} + \frac{\partial^2 g}{\partial^2 y}\right) = 0.$$

Together with the boundary condition in Equation (4) it determines the following Neumann's problem:

$$\begin{cases} \nabla^2 g = 0 & \text{in } D, \\ \frac{\partial g}{\partial n} = \frac{1}{2}\frac{\partial}{\partial s}\left(x^2 + y^2\right) & \text{on } \partial D, \end{cases} \tag{12}$$

with ∇^2 Laplace operator in the plane of the domain D. The solution $g(x,y)$ of this boundary value problem is unique up to an additive constant, and can be found in a closed form only for some simple cases (e.g., circular cross-sections, where $g = 0$ is an exact solution and fulfills the boundary condition [52]). Once the problem is solved, shear stresses are determined by means of Equation (11).

Another formulation of the problem can be derived defining *Prandtl's stress function* $\Phi(x,y)$ such that

$$\tau_{xz} = G\theta\frac{\partial \Phi}{\partial y}, \quad \tau_{yz} = -G\theta\frac{\partial \Phi}{\partial x}. \tag{13}$$

The equilibrium condition, Equation (9), is satisfied when the stress function Φ is a solution of the problem:

$$\begin{cases} \nabla^2 \Phi = -2 & \text{in } D, \\ d\Phi = 0 & \text{on } \partial D. \end{cases} \tag{14}$$

Since the addition of a constant to Φ does not affect stresses, the constant value Φ_0 on the outer boundary ∂D_0 can be set equal to 0. In hollow beams (multiple connected domains), the values of Φ_{0i} on the inner boundaries ∂D_i cannot be chosen arbitrarily, but must satisfy the circuital conditions

$$\int_{\partial D_i} \frac{\partial \Phi}{\partial n} \, ds = 2\Omega_i, \quad i = 1, 2, \ldots, n$$

where Ω_i denotes the area of the ith cavity. This constraint, which is necessary to ensure the displacement w has one value, cannot in general be solved explicitly within numerical analysis [56,63].

The torsion cross sectional rigidity J, entering the linear constitutive relationship for the twisting couple,

$$M_z = GJ\theta,$$

can be determined by integrating Prantdl's stress function over the section [52,54–58,62]:

$$J = 2 \int_D \Phi \, dxdy.$$

5. Circular Notched Cross-Section

As recalled in previous sections, the shear stress field is the solution of elliptic boundary value problems defined on the cross-section of the beam. Such problems are generally solved by means of finite differences or FEA.

However, the scheme presented here takes advantage of the simple form that the stress functions present for some particular geometries.

The warping function $g(x,y)$ and its harmonic conjugate $Z(x,y) = \Phi(x,y) + \frac{1}{2}(x^2 + y^2)$ are two functions which are coupled by the Cauchy-Riemann relations, i.e.,

$$\frac{\partial g}{\partial x} = \frac{\partial Z}{\partial y},$$

$$\frac{\partial g}{\partial y} = -\frac{\partial Z}{\partial x}.$$

Simple analytical solutions of Laplace equation are given by polynomials in the complex variable $x + iy$, such that the generic m-th term is

$$g + iZ = (p_m + iq_m)(x + iy)^m, \tag{15}$$

with [52]

$$g = p_m \left(x^m - \frac{m(m-1)}{2} x^{m-2} y^2 + \dots \right) +$$
$$- q_m \left(m x^{m-1} y + \dots \right),$$
$$Z = p_m \left(m x^{m-1} y + \dots \right) +$$
$$+ q_m \left(x^m - \frac{m(m-1)}{2} x^{m-2} y^2 + \dots \right).$$

In the case of negative values of m in Equation (15), the function $g + iZ$ will present *poles*, i.e., singularity points where the function approaches an infinite value. Thus, negative exponents m are possible in Equation (15), provided that the generated poles fall out of the elastic region D. For example, for $m = -1$, there is one simple pole in the origin:

$$g + iZ = \frac{(p_{-1} + iq_{-1})(x - iy)}{x^2 + y^2}.$$

In this case, the stress function

$$\Phi(x,y) = \frac{-hr^2 x}{x^2 + y^2} - \frac{1}{2}\left(x^2 + y^2\right) + hx + \frac{1}{2}r^2 \tag{16}$$

is regular inside the domain D, vanishes on the boundaries $x^2 + y^2 = r^2$ and $(x - h)^2 + y^2 = h^2$ and satisfies the problem in Equation (14) (it is Prandtl's stress function for the torsion of a circular cross-section beam, of radius h, with a circular notch of radius r centred on its surface, see Figure 6). Considering the details shown in Figure 4, the circular shape of notches is assumed to be a first approximation fitting the real profile.

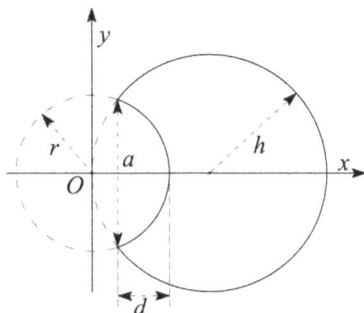

Figure 6. Reference frame, cross-section with radius h, and notch with radius r.

The associated warping function is, up to a constant,

$$g(x,y) = \left(1 - \frac{r^2}{x^2 + y^2}\right) h\, y. \tag{17}$$

Thus, for a beam with such a cross-section undergoing pure torsion, shear stresses can be easily evaluated with Equation (11) or (13) by partial differentiation of either g or Φ.

6. Theoretical Aspects in the Finite Element Formulation

The second order partial differential equation problem (with Dirichlet or Neumann boundary conditions) can be written, in its strong formulation, as:

$$\begin{cases} \nabla^2 \varphi = f_0(x,y) & \text{in } D \\ \dfrac{\partial \varphi}{\partial n} = f_1(x,y) \text{ or } \varphi(s) = f_2(s) & \text{in } \partial D \end{cases} \tag{18}$$

in the previous equations,

$$f_0(x,y) = 0,$$
$$f_1(x,y) = \frac{1}{2}\frac{\partial}{\partial s}\left(x^2 + y^2\right),$$

in the case $\varphi(x,y) = g$, while

$$f_0(x,y) = -2,$$
$$f_2(x,y) = \Phi_0,$$

in the case $\varphi(x,y) = \Phi$.

The weak form of the problem, which can be implemented by most finite element solvers, can be obtained by a weighted residuals method, introducing a set of weighting functions, η. The integral sum over the domain D in the product of the weights η by the residual $R = -\nabla^2 \varphi + f_0(x,y)$ is required to vanish

$$G(\varphi, \eta) = \int_D R\eta \, d\Omega = \int_D \left(-\nabla^2 \varphi + f_0\right) \eta \, d\Omega = 0. \tag{19}$$

Integration by parts, and Neumann's boundary condition, provide the weak form of the boundary value problem:

$$G(\varphi, \eta) = \int_D \nabla \varphi \cdot \nabla \eta + \eta f_0 \, d\Omega - \int_{\partial D} \eta f_1 \, ds = 0, \tag{20}$$

with ∇ the gradient in the plane of the domain D.

The condition for η is $\eta = 0$ on ∂D. The integral over the boundary will then vanish in the case of Dirichlet boundary conditions.

Equation (20) can be solved approximately using the finite element method; the domain D is first decomposed into a number ℓ of sub-domains, each one having n nodes.

The nodal approximations for the coordinates vector $\mathbf{x} = [x, y]^T$, the unknown function φ and the weighting function η are built using the same shape function N_I. Once their values in the nodes are known, the interpolations which derive are:

$$\tilde{\mathbf{x}} = \sum_{I=1}^{n} N_I \mathbf{x}_I,$$

$$\tilde{\varphi} = \sum_{I=1}^{n} N_I \varphi_I, \tag{21}$$

$$\tilde{\eta} = \sum_{I=1}^{n} N_I \eta_I,$$

where the ~ (tilde) sign denotes the value of the indicated function.

Inserting the nodal approximations in Equation (21) into the weak form of Equation (20), and summing over the ℓ subdomains leads to the finite element problem:

$$G(\varphi, \eta) = \sum_{e=1}^{\ell} G^e = \sum_{e=1}^{\ell} \sum_{I=1}^{n} \sum_{K=1}^{n} \eta_I \left(K_{IK}^e \varphi_K - P_I^e \right) = 0, \tag{22}$$

with the global matrix K_{IK} and the vector P_I formed by the elementary matrices and vectors:

$$K_{IK}^e = \int_{D^e} \left(\nabla N_I \cdot \nabla N_K \right) d\Omega^e \tag{23}$$

$$P_I^e = \int_{D^e} f_0(x, y) N_I \, d\Omega^e + \int_{\partial D} f_1(x, y) N_I \, ds. \tag{24}$$

To compute shear stresses, we must solve the linear system in Equation (22). To this end, the cross-section D properties, the second order moments of inertia, I_{xx}, I_{xy}, I_{yy}, and the centroid coordinates x_S and y_S, as well as the derivatives of the shape functions N with respect to x and y, must be known.

7. Stress Analysis

The presented methods can be specialized for the study of the above-mentioned class of CSFH-based microdevices. A member of this class is characterized by the presence of an embedded CSFH hinge, which is composed of a flexible curved element connecting a couple of conjugate profiles. One possible configuration is depicted in Figure 7. Relative motion among the pseudo-rigid parts i and j is possible because the thin cross-section of the curved element k behaves as a flexible rod.

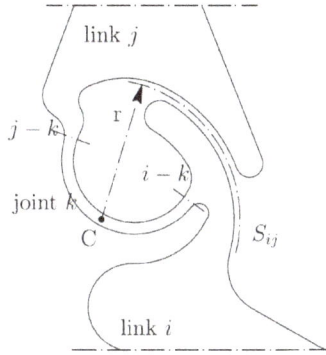

Figure 7. Flexural hinge geometry (see also Refs. [15,64]).

The technological limitations or the application complexity do not guarantee that the flexible curved beam *k* is subject only to *pure bending*. On the contrary, the tension field is rather complex and also includes shear component on the cross section. In fact, if a curved beam is considered, in-plane shear stress (as in Figure 2b) is always present, while out-of-plane, or *transverse*, shear stress (Figure 2c) is likely to be present for several reasons. One of them, already pointed out in Section 1, is that the shearing force on the beam is quite likely eccentric with respect to the shear center, thus implying a twisting couple. Considering that beams are thin-walled, we know this implies remarkable shear stress.

Since the aim of this paper is restricted to the analysis of stress concentration, it is reasonable to simplify the general case of *curved beam* by a *straight axis beam with in-plane shear* at one end. For this reason, the stress distribution over a cross section of the beam is investigated, and increase in the overall tension maximum value is monitored to avoid unpleasant damages or fractures at unexpectedly low loads.

In the following paragraphs, the elementary beam theory is applied to the case of a circular cross-section shaft with a circular shaped notch undergoing pure torsion. Stresses are computed by using the analytical solution of the equations shown in Section 5. Then, two more methods, both having numerical nature, are used to compare their results with the analytical solutions: finite differences and finite elements.

7.1. Analytical Solution

For the geometry in Figure 6, it has been shown that Prandtl's stress function

$$\Phi(x,y) = \frac{-hr^2x}{x^2+y^2} - \frac{1}{2}\left(x^2+y^2\right) + hx + \frac{1}{2}r^2,$$

and the warping function

$$g(x,y) = \left(1 - \frac{r^2}{x^2+y^2}\right)hy,$$

have to satisfy Equations (11) and (13):

$$\tau_{xz} = G\theta\left(\frac{\partial g}{\partial x} - y\right) = G\theta\frac{\partial\Phi}{\partial y},$$

$$\tau_{yz} = G\theta\left(\frac{\partial g}{\partial y} + x\right) = -G\theta\frac{\partial\Phi}{\partial x}.$$

The stress function and the partial derivatives can be rewritten in terms of polar coordinates, by introducing $\rho^2 = x^2 + y^2$:

$$\Phi = xh \left[1 - \left(\frac{r}{\rho} \right)^2 \right] - \frac{1}{2} \left(\rho^2 - r^2 \right),$$

$$\frac{\partial \Phi}{\partial x} = h \left[1 + \left(\frac{r}{\rho} \right)^2 \right] - x,$$

$$\frac{\partial \Phi}{\partial y} = y \left[2xh \left(\frac{r}{\rho} \right)^2 - 1 \right].$$

Some physical parameters have to be imposed to evaluate the vector $\tau = [\tau_{xz}, \tau_{yz}]^T$: the radii of the beam and of the notch h and r, as well as the twist of the centroidal fiber θ and the shear modulus G. Some plausible values are considered, although they do not resemble the exact geometry of the real device. This does not affect the accuracy of the results, since this calculations are not intended to give an exact solution, for which more accurate experimental studies are needed, in the first place. However, it is thought that, by assigning proper values to the parameters, the following computations give a close enough picture of the real stress field at a section of the hinge.

From the literature [32], we set 1 μm as fixed value for r, radius of the notch. Stresses were then studied for h linearly varying from 1 μm to 20 μm with steps of 1 μm.

We then imposed:

- beam radius, $h = 1 \div 20$ μm
- notch radius, $r = 1$ μm
- $G = 70$ GPa
- torsion relative angle $\omega_0 = 5° = 0.0873$ rad
- beam length $z = 300$ μm
- twist of the centroidal fiber $\theta = \frac{d\omega_0}{dz} = 291 \frac{\text{rad}}{\text{m}}$

The results were implemented and plotted using Matlab$^©$. The stresses were initially computed deriving partially Φ in a 10,000 nodes sample grid in x and y over the domain D; the grid was then refined with 40,000 and, finally, 160,000 nodes.

Figure 8a,b represents the shear vector fields in the cross section and around the notch for the two cases $h = 10r$ and $h = 20r$, respectively, in different scaled grids.

Figure 9a-c shows the shear stress field obtained by composing τ_{xz} and τ_{yz}, for different values of $\frac{h}{r}$ ratio, 20, 10 and 1, respectively.

However, for the sake of the present investigation, the area around the notch is more interesting than the bulk of the cross-section, and so a detailed computation was performed within the same coordinate system and number of nodes over a sample grid narrowed to a $3r \times 3r$ square. Results, for $h = 20r$, are shown in Figure 10a. Similar conclusions can be obtained by considering the case for which $h = r$, as depicted in Figure 10b.

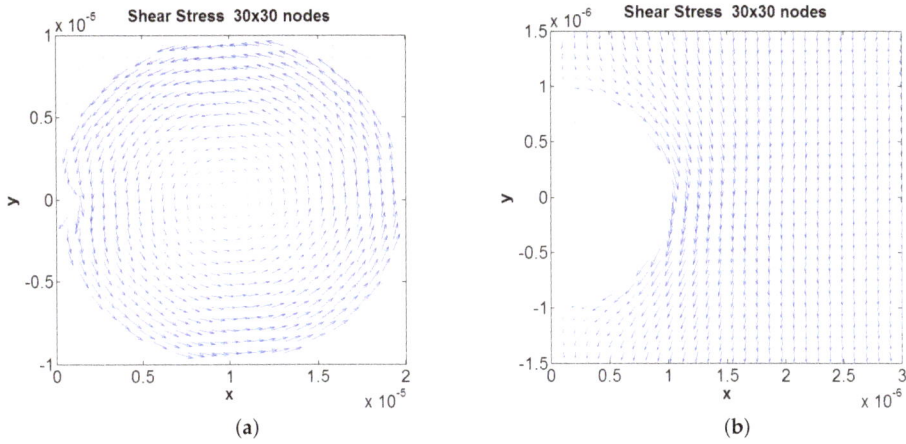

Figure 8. Shear stress vector field with the analytical approach in a: (**a**) 20 μm × 20 μm grid; and (**b**) 3 μm × 3 μm grid.

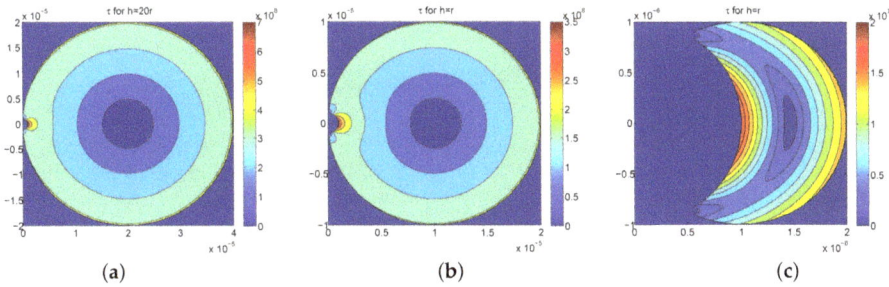

Figure 9. Shear stress module (Pa) in the cross section calculated through a 400×400 nodes grid with the analytical approach where: $\frac{h}{r} = 20$ (**a**); $\frac{h}{r} = 10$ (**b**); and $\frac{h}{r} = 20$ (**c**).

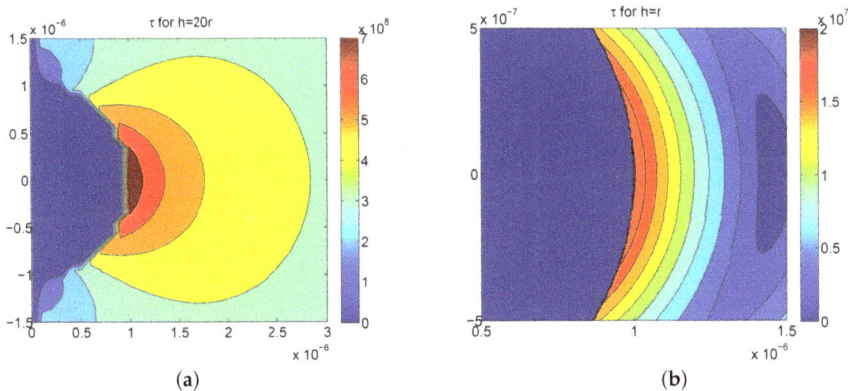

Figure 10. Shear stress module (Pa) around the notch for: $\frac{h}{r} = 20$ (**a**); and $\frac{h}{r} = 1$ (**b**).

Stress is concentrated around the circular arc midpoint, being the point where the maximum stress is attained.

7.2. Numerical Solution via Finite Difference Discretization

The computation by means of finite differences has been implemented using the free version of the software FlexPDE (*Version 6.36*).

The resulting stress field, obtained over a 587 nodes grid, is shown in Figure 11a–c, together with the plots of its two components τ_{xz} and τ_{yz}, and the stress function Φ.

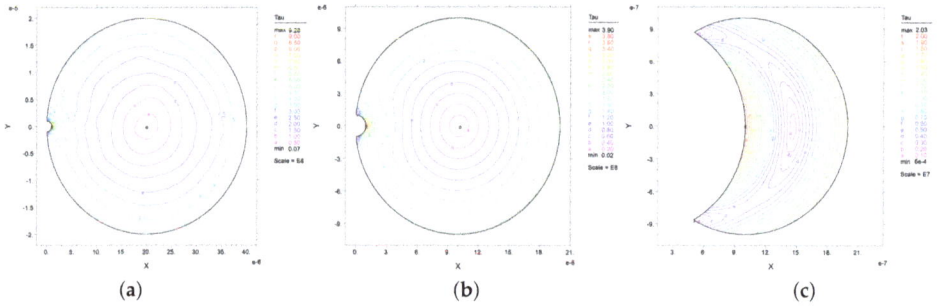

Figure 11. Shear stress (10^8 Pa) map for: $h = 20r$ (**a**); $h = 10r$ (**b**); and $h = r$ (**c**).

7.3. Numerical Solution via Finite Element Analysis

Finally, straight axis beams with circular section and circular notch were modeled by means of standard 3D modeling and FEA package ANSYS©.

Three different geometries were considered, thus three different solid models were built. Taking $r = 1$ μm, three cross sections with $h = 20r$, $h = 10r$ and $h = r$, respectively, were used to extrude the solid model of the beam. Using the same physical parameters in the previous methods, and imposing again 5° relative rotation among the end sections, the values of tension modules were obtained at the nodes.

The adopted meshes, for the three h to r ratios, are represented in Figures 12a, 13a and 14a.

The maximum shear stress fields are also reported in Figure 12b,c, for the case $h = r$, while the two groups in Figures 13b,c and 14b,c concern the two cases $h = 10r$ and $h = 20r$, respectively.

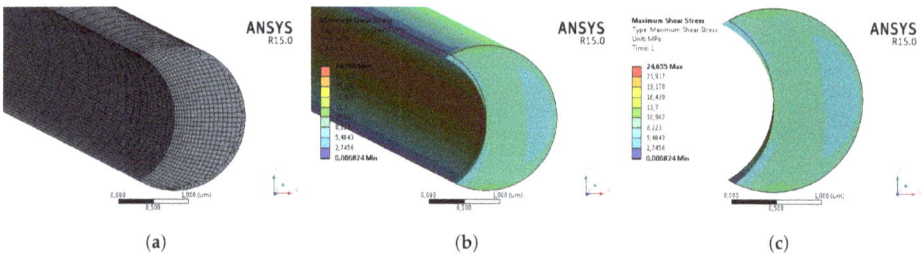

Figure 12. Summary of the results obtained by applying FEA to the case $r = h$: (**a**) mesh; (**b**) shear stress; and (**c**) shear stress in the section.

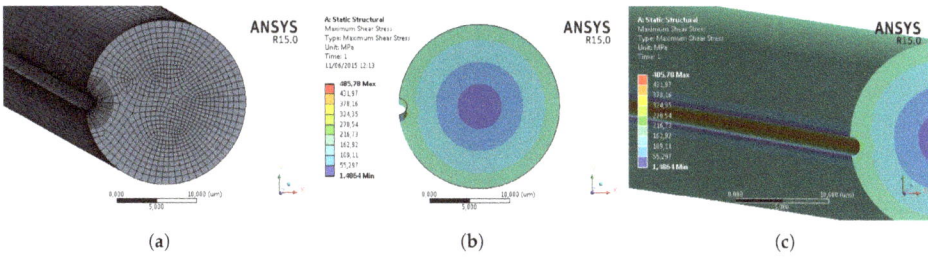

Figure 13. FEA applied to the case $h = 10r$: (**a**) mesh; and (**b,c**) shear stress.

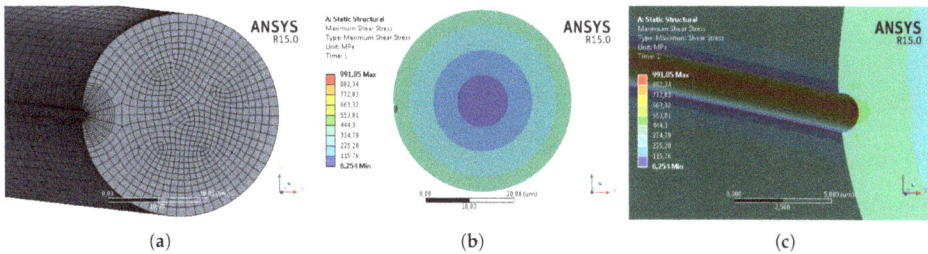

Figure 14. FEA applied to the case $h = 20r$: (**a**) mesh; and (**b,c**) shear stress.

7.4. Discussion about the Achieved Results

The obtained results have shown that the maximum shear stress holds in the center of the notch arc. Such point is called *core point* in Table 1. On the other hand, two other points have been considered, namely, the the upper and lower edge points of the notch, which correspond to the scallops spikes. In such points, called *edge points* in Table 1, the stress (denoted with τ_{edge}) is virtually null. This is consistent with the fact that such point belongs to a *stagnation zone* of the tension field.

Table 1. Comparison of Results.

h	Point	Position x	Position y	τ (MPa) Analytical	Flex PDE	FEA mid*	FEA min, max*
r	Core	r	0	19.9	20.3	20.37	24.66
	Edge	$\frac{r}{2}$	$\pm\frac{\sqrt{3}}{2}r$	0	$<6.0 \times 10^{-3}$	6.8×10^{-3}	6.8×10^{-3}
$10r$	Core	r	0	356	390	391.93	485.78
	Edge	$\frac{r}{20}$	$\pm\frac{\sqrt{399}}{20}r$	0	<20	1.49	1.49
$20r$	Core	r	0	727	928	802.01	991.85
	Edge	$\frac{r}{40}$	$\pm\frac{\sqrt{1599}}{40}r$	0	<150	7.08	6.25

mid: referred to mid section; min, max: minimum and maximum values.

Of course, symmetry implies identical solution for the top and the bottom edge points. Shear stress has been computed in these points using three methods. The analytical solution yields null values (i.e., zero, to the best of the numerical approximation of the floating point accuracy), while the other two methods give values much smaller (practically null values) than the those calculated in the core. This difference with the analytical method is certainly due to the discretization approximation of the numerical approaches.

The proposed analytical method gave us the opportunity to calculate stress in many different points and sections. For example, τ_{edge} and τ_{max} could be calculated as a function of the ratio ρ of the notch radius r to the circular section radius h. The results obtained have shown that the maximum stress decreases as $\rho = \frac{r}{h}$ increases. This is consistent with the physics of the phenomenon since, as h reduces in size, the discontinuity in the field becomes less severe. Considering the *edge points*, τ_{edge} also shows a similar decreasing trend as h decreases. As mentioned above, its magnitude remains significantly lower than the τ_{max} stress achieved at the notch core. Figure 15 shows the dependency of τ_{max} and τ_{edge} upon the $\frac{r}{h}$ ratio.

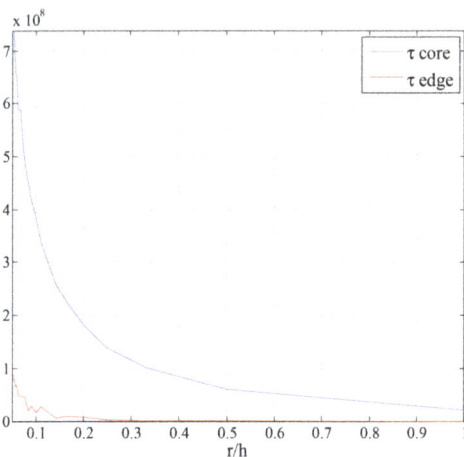

Figure 15. Stress (Pa) at the edge and at the core of the notch, as a function of $\rho = \frac{r}{h} r$.

The diagram reported in Figure 15 concerns the tension absolute values that hold in the notch core and edge as a function of ρ. However, this result would convey rather little insight if it was not reduced to a physical model *equivalent* to the scalloped notch. For this purpose, there are still two particular problems to be solved:

- the definition of an equivalence criterion for relating the adopted simplified geometry (see Figure 6) to the original geometry (see Figures 2 and 4); and
- the introduction of a reliable index of the severity of stress concentration.

7.4.1. Adopted Equivalence Criterion

One advantage of the adopted cross section is that it is very simply described by means of analytical and geometrical tools. For example, referring back again to Figure 6, it is clear that the scalloped notch is represented by the circular segment having r, d and a as radius, depth and chord, respectively. One characteristic of scalloping is the dimensionless ratio δ of the notch depth d to the chord a. The two dimensionless parameters ρ and δ can be easily related using two simple, well-known geometrical equations for the circular segment, namely,

$$d = r - \sqrt{r^2 - \frac{a^2}{4}}$$
$$a = \frac{r\sqrt{4h^2 - r^2}}{h} \tag{25}$$

7.4.2. Stress Concentration Factor

The *stress concentration factor* K_t, for the considered beam, can be calculated as the ratio of the maximum stress on the notch core $\tau_{\max}$ by the stress calculated on a point positioned on the other side of the beam section (with reference to Figure 6, this points is on the x axis, opposite to the origin O).

The adopted equivalence criterion and the introduction of K_t give rise to an interesting interpretation of the iterative application of the analytical method.

Figure 16 reports the values of the stress concentration factor K_t as a function of the parameter $\eta = \rho^{-1} = \frac{h}{r}$, which spans, in the real domain, from $\frac{1}{2}$ (which correspond to a beam with null cross sectional area) to ∞. There are no reasons for η to go beyond the order of magnitude of tens and so 40 has been regarded as the maximum. The trend shows a sudden increase of K_t for small increments of η. For values of η greater than about 5, K_t tends asymptotically to 2. However, according to the equivalence criterion, the scalloping geometry is better characterized by the ratio $\delta = \frac{d}{a}$.

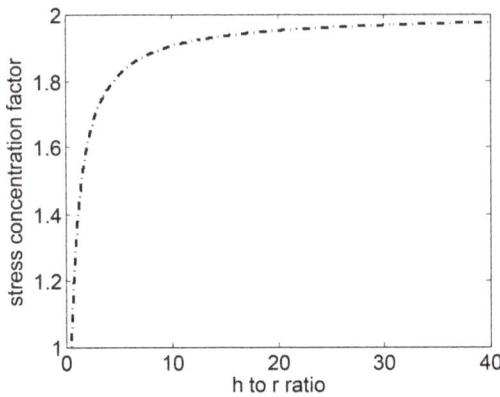

Figure 16. Stress Concentration Factor vs. $\eta = \frac{h}{r}$.

By using Equations (25), it is possible to evaluate the ratio δ of the scalloping dept d to chord a as a function of η (see Figure 17).

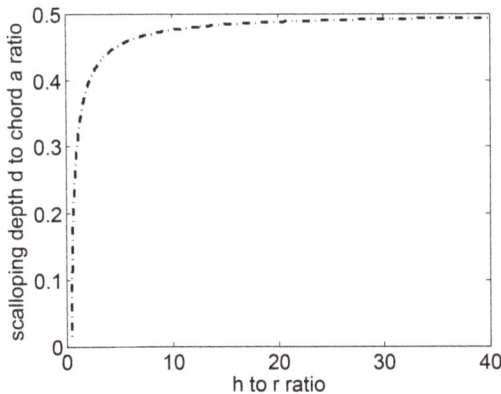

Figure 17. Ratio δ of the scalloping dept d to chord a as a function of η.

Finally, the relation between η and δ can be used to obtain the explicit dependency of K_t upon δ, as reported in Figure 18. In fact, the dependency of K_t on η and of the latter on δ, can be combined together to get, via function composition, $K_t(\delta)$.

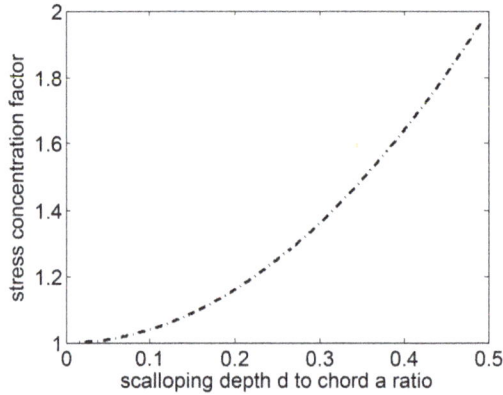

Figure 18. Stress Concentration Factor vs. δ.

When $\delta = 0$, which corresponds to a smooth wall, no stress concentration occurs ($K_t = 1$). However, for positive values of δ, the trend depicted in Figure 18 shows that the notch depth d has a great impact on stress concentration. The maximum value of K_t is 2 and it corresponds to a notch which has its center on a vertical wall (with $h \gg r$).

Thanks to the above-presented analysis, it is possible to state that the stress concentration factor tends asymptotically to the theoretical value 2 as the scalloping depth increases with respect to the scalloped chord, where the limit value is already known in the literature (see, for example, Ref. [65]).

Consequently, it can be also stated, as a general principle, that, for a DRIE micromachined beam with apparent and not treated scalloping, the strength can decrease to the half value of the strength of the same beam with no notches.

8. Conclusions

The wide range of applications has made MEMS more complicated. Therefore, their mechanical structure and internal stress state also become more complex. In many of those applications, MEMS are called to interact with the environment and so they need high aspect ratios, which can be achieved by manufacturing them via DRIE process. Unfortunately, the etched profiles cannot always get rid of the scalloping problem, which induces a dangerous increase of concentrated stress in the structure. Since this stress concentration gives rise to evident loss of structural strength for silicon, it is important to take into account the effects of scalloping on stress concentration in this kind of structures.

In this paper, three different methods are proposed for stress concentration characterization of silicon DRIE micromachined structures. All of them use a straight axis beam with a circular section and a circular notch. The first approach is based on the theory of linear elasticity and evaluates the shear stresses for pure torsion (stress concentration being due to shear state only). The other two methods use numerical approach: finite difference and finite element analysis. Although the three methods are very different in nature, the obtained values are fairly comparable. Furthermore, using the analytical method makes it easier to obtain diagrams that show the stress concentration factor with respect to the notch radius. The asymptotic value of the stress concentration factor, as a function of the notch relative radius, is in accordance with the literature. This leads to the conclusion that, in DRIE manufactured notched beams, strength is virtually half of the one with no notches.

Author Contributions: Conceptualization, all authors; Methodology, S.B. and G.R.; Software, S.B. and N.P.B.; Validation, all authors; Formal Analysis, S.B., G.R. and M.V.; Investigation and resources for fabrication, A.B. and P.B.; Data curation, all authors; Writing-original draft preparation, review, editing, and visualization, M.V. and N.P.B.; and Supervision, P.B. and G.R.

Funding: This research received no external funding.

Conflicts of Interest: The authors declare no conflict of interest.

References

1. Elman, N.; Ho Duc, H.; Cima, M. An implantable MEMS drug delivery device for rapid delivery in ambulatory emergency care. *Biomed. Microdevices* **2009**, *11*, 625–631. [CrossRef] [PubMed]
2. Li, Y.; Shawgo, R.; Tyler, B.; Henderson, P.; Vogel, J.; Rosenberg, A.; Storm, P.; Langer, R.; Brem, H.; Cima, M. In vivo release from a drug delivery MEMS device. *J. Control. Release* **2004**, *100*, 211–219. [CrossRef] [PubMed]
3. Lo, R.; Li, P.Y.; Saati, S.; Agrawal, R.; Humayun, M.; Meng, E. A passive MEMS drug delivery pump for treatment of ocular diseases. *Biomed. Microdevices* **2009**, *11*, 959–970. [CrossRef] [PubMed]
4. Voskerician, G.; Shive, M.; Shawgo, R.; Von Recum, H.; Anderson, J.; Cima, M.; Langer, R. Biocompatibility and biofouling of MEMS drug delivery devices. *Biomaterials* **2003**, *24*, 1959–1967. [CrossRef]
5. Xie, Y.; Xu, B.; Gao, Y. Controlled transdermal delivery of model drug compounds by MEMS microneedle array. *Nanomed. Nanotechnol. Biol. Med.* **2005**, *1*, 184–190. [CrossRef] [PubMed]
6. Ibrahim, D. Using MEMS accelerometers in aerospace and defence electronics. *Electron. World* **2012**, *118*, 16–21.
7. Leclerc, J. MEMs for aerospace navigation. *IEEE Aerosp. Electron. Syst. Mag.* **2007**, *22*, 31–36. [CrossRef]
8. Hsu, H.; Hariz, A.; Omari, T.; Teng, M.; Sii, D.; Chan, S.; Lau, L.; Tan, S.; Lin, G.; Haskard, M.; et al. Development of a MEMS based manometric catheter for diagnosis of functional swallowing disorders. *J. Phys. Conf. Ser.* **2006**, *34*, 955–960. [CrossRef]
9. Sun, H.; Fu, G.; Xie, H. A MEMS accelerometer-based real-time motion-sensing module for urological diagnosis and treatment. *J. Med. Eng. Technol.* **2013**, *37*, 127–134. [CrossRef] [PubMed]
10. Horvath, I.; Panayotatos, P.; Lu, Y. Si MEMS microbearing with integrated safety sensors for surgical applications. *Microelectron. J.* **2001**, *32*, 1–9. [CrossRef]
11. Lou, L.; Ramakrishna, K.; Shao, L.; Park, W.; Yu, D.; Lim, L.; Wee, Y.; Kripesh, V.; Feng, H.; Chua, B.; et al. Sensorized guidewires with MEMS tri-axial force sensor for minimally invasive surgical applications. In Proceedings of the 2010 Annual International Conference of the IEEE Engineering in Medicine and Biology, Buenos Aires, Argentina, 31 August–4 September 2010; pp. 6461–6464.
12. To, G.; Qu, W.; Mahfouz, M. ASIC Design for Wireless Surgical MEMS Device and Instrumentation. In Proceedings of the 2006 International Conference of the IEEE Engineering in Medicine and Biology Society, New York, NY, USA, 30 August–3 September 2006; pp. 5892–5895.
13. Chronis, N.; Lee, L. Polymer mems-based microgripper for single cell manipulation. In Proceedings of the 17th IEEE International Conference on Micro Electro Mechanical Systems (MEMS), Maastricht, The Netherlands, 25–29 January 2004; pp. 17–20.
14. Kim, K.; Liu, X.; Zhang, Y.; Sun, Y. Nanonewton force-controlled manipulation of biological cells using a monolithic MEMS microgripper with two-axis force feedback. *J. Micromech. Microeng.* **2008**, *18*, 055013. [CrossRef]
15. Belfiore, N.P.; Verotti, M.; Crescenzi, R.; Balucani, M. Design, optimization and construction of MEMS-based micro grippers for cell manipulation. In Proceedings of the ICSSE 2013—IEEE International Conference on System Science and Engineering, Budapest, Hungary, 4–6 July 2013; pp. 105–110.
16. Balucani, M.; Belfiore, N.P.; Crescenzi, R.; Verotti, M. The development of a MEMS/NEMS-based 3 D.O.F. compliant micro robot. *Int. J. Mech. Control* **2011**, *12*, 3–10.
17. Belfiore, N.P.; Simeone, P. Inverse kinetostatic analysis of compliant four-bar linkages. *Mech. Mach. Theory* **2013**, *69*, 350–372. [CrossRef]
18. Verotti, M.; Dochshanov, A.; Belfiore, N.P. Compliance Synthesis of CSFH MEMS-Based Microgrippers. *J. Mech. Des.-Trans. ASME* **2017**, *139*, 022301. [CrossRef]

19. Belfiore, N.P.; Broggiato, G.; Verotti, M.; Crescenzi, R.; Balucani, M.; Bagolini, A.; Bellutti, P.; Boscardin, M. Development of a MEMS technology CSFH based microgripper. In Proceedings of the 23rd International Conference on Robotics in Alpe-Adria-Danube Region, Smolenice, Slovakia, 3–5 September 2014; Institute of Electrical and Electronics Engineers Inc.: Piscataway Township, NJ, USA, 2015.

20. Bagolini, A.; Ronchin, S.; Bellutti, P.; Chistè, M.; Verotti, M.; Belfiore, N.P. Fabrication of Novel MEMS Microgrippers by Deep Reactive Ion Etching With Metal Hard Mask. *J. Microelectromech. Syst.* **2017**, *26*, 926–934. [CrossRef]

21. Belfiore, N.P.; Scaccia, M.; Ianniello, F.; Presta, M. Selective Compliance Hinge. U.S. Patent 8,191,204 B2, 5 June 2012.

22. Balucani, M.; Belfiore, N.P.; Crescenzi, R.; Genua, M.; Verotti, M. Developing and modeling a plane 3 DOF compliant micromanipulator by means of a dedicated MBS code. In Proceedings of the 2011 NSTI Nanotechnology Conference and Expo, NSTI-Nanotech, At Boston, MA, USA, 13–16 June 2011; Volume 2, pp. 659–662.

23. Belfiore, N.P.; Balucani, M.; Crescenzi, R.; Verotti, M. Performance analysis of compliant mems parallel robots through pseudo-rigid-body model synthesis. In Proceedings of the ASME 2012 11th Biennial Conference on Engineering Systems Design and Analysis, Nantes, France, 2–4 July 2012; Volume 3, pp. 329–334.

24. Belfiore, N.P.; Emamimeibodi, M.; Verotti, M.; Crescenzi, R.; Balucani, M.; Nenzi, P. Kinetostatic optimization of a MEMS-based compliant 3 DOF plane parallel platform. In Proceedings of the ICCC 2013—IEEE 9th International Conference on Computational Cybernetics, Tihany, Hungary, 8–10 July 2013; pp. 261–266.

25. Nenzi, P.; Crescenzi, R.; Dolgyi, A.; Klyshko, A.; Bondarenko, V.; Belfiore, N.P.; Balucani, M. High density compliant contacting technology for integrated high power modules in automotive applications. In Proceedings of the Electronic Components and Technology Conference, San Diego, CA, USA, 29 May–1 June 2012; pp. 1976–1983.

26. Dochshanov, A.; Verotti, M.; Belfiore, N.P. A Comprehensive Survey on Microgrippers Design: Operational Strategy. *J. Mech. Des.-Trans. ASME* **2017**, *139*, 070801. [CrossRef]

27. Verotti, M.; Dochshanov, A.; Belfiore, N.P. A Comprehensive Survey on Microgrippers Design: Mechanical Structure. *J. Mech. Des.-Trans. ASME* **2017**, *139*, 060801. [CrossRef]

28. Di Giamberardino, P.; Bagolini, A.; Bellutti, P.; Rudas, I.; Verotti, M.; Botta, F.; Belfiore, N.P. New MEMS tweezers for the viscoelastic characterization of soft materials at the microscale. *Micromachines* **2017**, *9*, 15. [CrossRef]

29. Belfiore, N.P.; Prosperi, G.; Crescenzi, R. A simple application of conjugate profile theory to the development of a silicon micro tribometer. In Proceedings of the ASME 2014 12th Biennial Conference on Engineering Systems Design and Analysis, Copenhagen, Denmark, 25–27 July 2014; Volume 2.

30. Laermer, F.; Schlip, A. Method of Anisotropically Etching Silicon. U.S. Patent 5,501,893, 26 March 1996.

31. Ayön, A.A.; Chen, K.S.; Lohner, K.A.; Spearing, S.M.; Sawin, H.H.; Schmidt, M.A. Deep Reactive Ion Etching of Silicon. In Proceedings of the Symposium AA Materials Science of Microelectromechanical Systems (MEMS), Boston, MA, USA, 1–2 December 1998; Volume 546.

32. Chen, K.S.; Ayön, A.A.; Zhang, X.; Spearing, S. Effect of process parameters on the surface morphology and mechanical performance of silicon structures after deep reactive ion etching (DRIE). *J. Microelectromech. Syst.* **2002**, *11*, 264–275. [CrossRef]

33. Fricker, H. Why does charge concentrate on points? *Phys. Educ.* **1989**, *24*, 157. [CrossRef]

34. Crescenzi, R.; Balucani, M.; Belfiore, N.P. Operational characterization of CSFH MEMS technology based hinges. *J. Micromech. Microeng.* **2018**, *28*, 055012. [CrossRef]

35. Mita, Y.; Sugiyama, M.; Kubota, M.; Marty, F.; Bourouina, T.; Shibata, T. Aspect Ratio Dependent Scalloping Attenuation in Drie and an Application to Low-Loss Fiber-Optical Switches. In Proceedings of the 19th IEEE International Conference on Micro Electro Mechanical Systems, Istanbul, Turkey, 22–26 January 2006; pp. 114–117.

36. Pham, P.; Dao, D.; Amaya, S.; Kitada, R.; Sugiyama, S. Fabrication and characterization of smooth Si mold for hot embossing process. *IEEJ Trans. Sens. Micromach.* **2007**, *127*, 187–191. [CrossRef]

37. Song, I.H.; Peter, Y.A.; Meunier, M. Smoothing dry-etched microstructure sidewalls using focused ion beam milling for optical applications. *J. Micromech. Microeng.* **2007**, *17*, 1593–1597. [CrossRef]

38. Defforge, T.; Song, X.; Gautier, G.; Tillocher, T.; Dussart, R.; Kouassi, S.; Tran-Van, F. Scalloping removal on DRIE via using low concentrated alkaline solutions at low temperature. *Sens. Actuators A-Phys.* **2011**, *170*, 114–120. [CrossRef]

39. Hung, Y.J.; Lee, S.L.; Thibeault, B.J.; Coldren, L.A. Realization of silicon nanopillar arrays with controllable sidewall profiles by holography lithography and a novel single-step deep reactive ion etching. In *Symposia P Q R Low-Dimensional Functional Nanostructures-Fabrication, Characterization and Applications*; Materials Research Society Online Proceedings Library Archive, Cambridge University Press: Cambridge, UK, 2010; Volume 1258.

40. Inagaki, N.; Sasaki, H.; Shikida, M.; Sato, K. Selective removal of micro-corrugation by anisotropic wet etching. In Proceedings of the TRANSDUCERS 2009—2009 International Solid-State Sensors, Actuators and Microsystems Conference, Denver, CO, USA, 21–25 June 2009; pp. 1865–1868.

41. Phinney, L.; McKenzie, B.; Ohlhausen, J.; Buchheit, T.; Shul, R. Characterization of SOI MEMS sidewall roughness. In Proceedings of the ASME 2011 International Mechanical Engineering Congress and Exposition, Denver, CO, USA, 11–17 November 2011; Volume 11, pp. 187–193.

42. Chekurov, N.; Grigoras, K.; Peltonen, A.; Franssila, S.; Tittonen, I. The fabrication of silicon nanostructures by local gallium implantation and cryogenic deep reactive ion etching. *Nanotechnology* **2009**, *20*, 065307. [CrossRef] [PubMed]

43. De Boer, M.; Gardeniers, J.; Jansen, H.; Smulders, E.; Gilde, M.J.; Roelofs, G.; Sasserath, J.; Elwenspoek, M. Guidelines for etching silicon MEMS structures using fluorine high-density plasmas at cryogenic temperatures. *J. Microelectromech. Syst.* **2002**, *11*, 385–401. [CrossRef]

44. Murakami, K.; Wakabayashi, Y.; Minami, K.; Esashi, M. Cryogenic dry etching for high aspect ratio microstructures. In Proceedings of the IEEE Micro Electro Mechanical Systems, Fort Lauderdale, FL, USA, 10 February 1993; IEEE: Piscataway, NJ, USA, 1993; pp. 65–70.

45. Sainiemi, L.; Franssila, S. Mask material effects in cryogenic deep reactive ion etching. *J. Vac. Sci. Technol. B* **2007**, *25*, 801–807. [CrossRef]

46. Morikawa, Y.; Murayama, T.; Sakuishi, T.; Nakamura, T.; Kurimoto, T.; Nakamuta, Y.; Kimura, I.; Suu, K. *Scallop Free si Etching and Low Cost Integration Technologies for 2.5D Si Interposer*; International Microelectronics Assembly and Packaging Society: Research Triangle Park, NC, USA, 2012; pp. 998–1000.

47. Guo, M.; Chou, X.; Mu, J.; Liu, B.; Xiong, J. Fabrication of micro-trench structures with high aspect ratio based on DRIE process for MEMS device applications. *Microsyst. Technol.* **2013**, *19*, 1097–1103. [CrossRef]

48. Miller, K.; Li, M.; Walsh, K.; Fu, X.A. The effects of DRIE operational parameters on vertically aligned micropillar arrays. *J. Micromech. Microeng.* **2013**, *23*, 035039. [CrossRef]

49. Ma, Z.; Jiang, C.; Yuan, W.; He, Y. Large-scale patterning of hydrophobic silicon nanostructure arrays fabricated by dual lithography and deep reactive ion etching. *Nano-Micro Lett.* **2013**, *5*, 7–12. [CrossRef]

50. Wang, Z.; Jiang, F.; Yu, D.; Zhang, W. Si Etching for TSV Formation. *ECS Trans.* **2014**, *60*, 407–412. [CrossRef]

51. Zhao, Y.P.; Wang, L.S.; Yu, T.X. Mechanics of adhesion in MEMS—A review. *J. Adhes. Sci. Technol.* **2003**, *17*, 519–546. [CrossRef]

52. Fraeijs de Veubeke, B.M. *A Course in Elasticity*; Springer: New York, NY, USA, 1979; pp. 135–200.

53. Boresi, A.; Schmidt, R. *Advanced Mechanics of Materials*; John Wiley & Sons: Hoboken, NJ, USA, 2003.

54. Dell'Isola, F.; Ruta, G.C. Outlooks in Saint-Venant theory III. Torsion and flexure in sections of variable thickness by formal expansions. *Arch. Mech.* **1997**, *49*, 321–343.

55. Andreaus, U.; Ruta, G. A review of the problem of the shear centre(s). *Contin. Mech. Thermodyn.* **1998**, *10*, 369–380. [CrossRef]

56. Paolone, A.; Ruta, G.; Vidoli, S. Torsion in multi-cell thin-walled girders. *Acta Mech.* **2009**, *206*, 163–171. [CrossRef]

57. Ruta, G. On the flexure of a Saint-Venant cylinder. *J. Elast.* **1998**, *52*, 99–110. [CrossRef]

58. Ruta, G. On Kelvin's formula for torsion of thin cylinders. *Mech. Res. Commun.* **1999**, *26*, 591–596. [CrossRef]

59. Saint-Venant, B. Mémoire sur la torsion des prismes. *Mem. Savants Etrang.* **1855**, *14*, 233–560.

60. Clebsch, R.F.A. *Theorie der Elasticität Fester Körper*; B. G. Teubner: Leipzig, Germany, 1862.

61. Iesan, D. On Saint-Venant's problem. *Arch. Ration. Mech. Anal.* **1986**, *91*, 363–373. [CrossRef]

62. Muskhelishvili, N.I. *Some Basic Problems of the Mathematical Theory of Elasticity*; Noordhoff Ltd.: Groningen, The Netherlands, 1963; pp. 571–607.

63. Gruttman, F.; Sauer, R.; Wagner, W. Shear Stresses in Prismatic Beams with Arbitrary Cross-Sections. *Int. J. Numer. Methods Eng.* **1999**, *45*, 865–889. [CrossRef]

64. Cecchi, R.; Verotti, M.; Capata, R.; Dochshanov, A.; Broggiato, G.; Crescenzi, R.; Balucani, M.; Natali, S.; Razzano, G.; Lucchese, F.; et al. Development of micro-grippers for tissue and cell manipulation with direct morphological comparison. *Micromachines* **2015**, *6*, 1710–1728. [CrossRef]

65. Timoshenko, S.; Goodier, J. *Theory of Elasticity*; McGraw-Hill: New York, NY, USA, 1970.

actuators

MDPI

Article

The Effects of Structure Thickness, Air Gap Thickness and Silicon Type on the Performance of a Horizontal Electrothermal MEMS Microgripper

Marija Cauchi [1,*], Ivan Grech [2], Bertram Mallia [3], Pierluigi Mollicone [1] and Nicholas Sammut [2]

[1] Department of Mechanical Engineering, Faculty of Engineering, University of Malta, MSD 2080 Msida, Malta; pierluigi.mollicone@um.edu.mt
[2] Department of Microelectronics and Nanoelectronics, Faculty of Information and Communication Technology, University of Malta, MSD 2080 Msida, Malta; ivan.grech@um.edu.mt (I.G.); nicholas.sammut@um.edu.mt (N.S.)
[3] Department of Metallurgy and Materials Engineering, Faculty of Engineering, University of Malta, MSD 2080 Msida, Malta; bertram.mallia@um.edu.mt
* Correspondence: mcauc03@um.edu.mt; Tel.: +356-2340-2023

Received: 20 June 2018; Accepted: 11 July 2018; Published: 15 July 2018

Abstract: The ongoing development of microelectromechanical systems (MEMS) over the past decades has made possible the achievement of high-precision micromanipulation within the micromanufacturing, microassembly and biomedical fields. This paper presents different design variants of a horizontal electrothermally actuated MEMS microgripper that are developed as microsystems to micromanipulate and study the deformability properties of human red blood cells (RBCs). The presented microgripper design variants are all based on the U-shape 'hot and cold arm' actuator configuration, and are fabricated using the commercially available Multi-User MEMS Processes (MUMPs®) that are produced by MEMSCAP, Inc. (Durham, NC, USA) and that include both surface micromachined (PolyMUMPs™) and silicon-on-insulator (SOIMUMPs™) MEMS fabrication technologies. The studied microgripper design variants have the same in-plane geometry, with their main differences arising from the thickness of the fabricated structures, the consequent air gap separation between the structure and the substrate surface, as well as the intrinsic nature of the silicon material used. These factors are all inherent characteristics of the specific fabrication technologies used. PolyMUMPs™ utilises polycrystalline silicon structures that are composed of two free-standing, independently stackable structural layers, enabling the user to achieve structure thicknesses of 1.5 μm, 2 μm and 3.5 μm, respectively, whereas SOIMUMPs™ utilises a 25 μm thick single crystal silicon structure having only one free-standing structural layer. The microgripper design variants are presented and compared in this work to investigate the effect of their differences on the temperature distribution and the achieved end-effector displacement. These design variants were analytically studied, as well as numerically modelled using finite element analysis where coupled electrothermomechanical simulations were carried out in CoventorWare® (Version 10, Coventor, Inc., Cary, NC, USA). Experimental results for the microgrippers' actuation under atmospheric pressure were obtained via optical microscopy studies for the PolyMUMPs™ structures, and they were found to be conforming with the predictions of the analytical and numerical models. The focus of this work is to identify which one of the studied design variants best optimises the microgripper's electrothermomechanical performance in terms of a sufficient lateral tip displacement, minimum out-of-plane displacement at the arm tips and good heat transfer to limit the temperature at the cell gripping zone, as required for the deformability study of RBCs.

Keywords: MEMS microgrippers; micromanipulation; red blood cells; electrothermal actuation; PolyMUMPs™; SOIMUMPs™; structure thickness; air gap thickness; polysilicon; single crystal silicon

1. Introduction

The constantly growing need for high-precision micromanipulation has seen the implementation of microelectromechanical systems (MEMS) in numerous applications within the micromanufacturing and biomedical fields. Microgrippers are one such type of MEMS devices that are typically designed to safely manipulate biological cells [1–9] as well as micromechanical parts [10–12]. The successful development of a MEMS microgripper depends on taking into account several factors including the design and mechanical structure [13], the actuation mechanism [14,15], the choice of materials used [16–19], the operational requirements and limitations [20], the fabrication technology [21–26], and the environment in which the microgripper will be operated [27].

Our previous work [25] presented an analytical and a numerical model of a horizontal electrothermal microgripper design that were developed to reliably predict the temperature distribution and displacement at the arm tips when operated under atmospheric pressure. These developed models focused on an electrothermomechanical analysis which considered heat conduction to the substrate, through both the air gap and the anchor pads, as the main source of heat loss from the structure. Tip displacement results obtained from the analytical and numerical models were comparable with the results obtained from the experimental testing performed on a fabricated PolyMUMPs™ structure. The main contribution of the current work consists of utilising the models established and validated in [25] to evaluate and compare the temperature and displacement behaviour of four electrothermal microgripper design variants. These microgripper design variants have the same in-plane geometry, with their main differences being the device thickness, the air gap thickness between the structure and the substrate surface, and the intrinsic nature of the silicon material, all of which are characteristics of the fabrication technologies used. In this regard, the current study investigates the feasibility of two commercially available fabrication technologies—PolyMUMPs™ and SOIMUMPs™—to realise a microgripper structure with optimised electrothermomechanical performance. The electrothermomechanical performance is optimised based on the requirements to micromanipulate and study the deformability properties of human red blood cells (RBCs), and focuses on obtaining a sufficient opening stroke of the microgripper arms, minimum out-of-plane displacement at the arm tips and minimum temperature rise at the cell gripping zone.

RBCs, with a diameter of around 8 μm and an approximate thickness of 2 μm, have an average lifetime of 120 days. They are responsible for carrying oxygen, and carbon dioxide (CO_2), from the lungs to the body tissues and back, respectively. Healthy and diseased RBCs differ in their deformability properties. Healthy RBCs are capable of undergoing the necessary cellular deformation to flow even through human microcirculation vessels with small diameters, thus being able to continuously fulfill their function of supplying oxygen to the biological tissues. However, studies have shown that a number of pathological conditions such as diabetes, sickle cell anemia, malaria and other genetic disorders negatively impact the deformability properties of RBCs [28,29]. The increased cell membrane stiffness in the presence of such pathological conditions inhibits the cellular deformation required for RBCs to freely flow through the smaller blood vessels, thus preventing RBCs from performing their important function. In the case of healthy cells, the required gripping force will be of the order of a few microNewtons, with this force expected to increase to achieve the same cell deformation in diseased cells due to the increased membrane stiffness. The deformability properties of RBCs can thus be an important measure of human well-being and their measurement has in fact been the subject of various studies over the years [30,31].

This paper investigates the effects of the structure thickness, the air gap thickness, and the different sources of silicon on the performance of a horizontal electrothermal microgripper designed for the deformability study of RBCs. The implemented 'hot and cold arm' microgripper design variants and the electrothermal actuation principle are defined in Section 2. Section 3 presents and compares the commercially available surface and bulk micromachining processes whose feasibility is studied for the fabrication of the considered microgripper design variants. The developed microgripper numerical models were simulated based on the finite element analysis (FEA) modelling technique and are described in Section 4. Section 5 outlines the setup used for the experimental investigation of the fabricated microgrippers that was carried out using optical microscopy. Section 6 presents and compares the electrical, thermal and mechanical performance of the different microgripper design variants, and finally, Section 7 highlights the concluding remarks and outcomes of this work.

2. Microgripper Design Variants and Principle of Operation

The microgripper design variants developed in this study are all derived from the U-shape 'hot and cold arm' electrothermal actuator configuration [32]. A 'hot and cold arm' actuator design is mainly composed of two parallel arms with dissimilar widths, the hot arm and the cold arm, that underlie the operating principle of the electrothermal actuator, and a flexure component that allows the actuator to bend and to achieve the required motion. Figure 1 demonstrates how two such actuators can form a complete microgripper, with the extending microgripper arms amplifying the achieved tip displacement. The 'hot and cold arm' microgripper design considered here consists of a kinematically simple structure that can be manufactured with a limited number of fabrication steps. Such a structure makes it easy to assess the feasibility of two commercially available fabrication processes as well as to study the effects of basic design parameters such as structure thickness, air gap thickness and silicon type on the performance of a microgripper for the intended application of RBC manipulation.

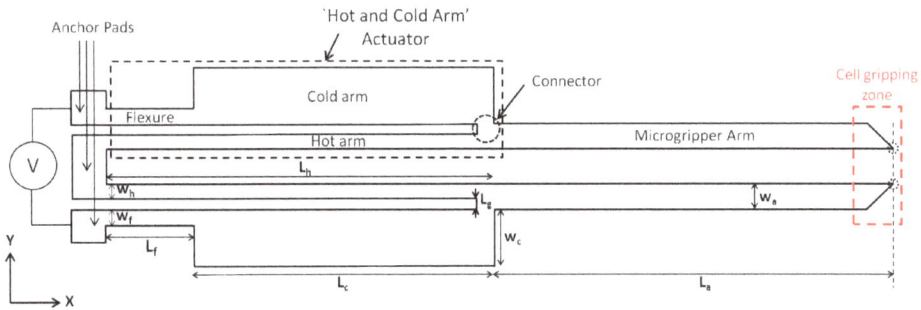

Figure 1. Schematic illustration of the electrothermal microgripper design based on the U-shape 'hot and cold arm' actuator configuration. The encircled points within the cell gripping zone indicate the arm tip points that will be later investigated for their *x*-, *y*- and *z*-displacements. The cell gripping zone is to undergo further development in future work.

Under an applied potential, the hot and cold arms experience unequal resistive heating due to their different cross-sectional areas, and the resulting asymmetric thermal expansions result in a bending moment and in angular rotation of the arms about the fixed anchor pads. This motion drives the two gripping arms to open and close as necessary, and results in non-parallel gripping of the micro-object. The relative rotation angle for the opening motion of the arms is given by the inverse tangent of the ratio of the arm tip's displacement to the combined length of the hot arm and the microgripper arm ($L_h + L_a$). In scenarios where parallel gripping is essential, a different kinematic design will be required [33]. There is, however, no requirement for non-parallel gripping in the case of RBC manipulation as RBCs are highly deformable and thus this non-parallel gripping is not expected to hinder their manipulation and deformability characterisation. Moreover, the same designed microgripper will be used to comparatively study the deformability properties of different RBCs, and hence the non-parallel gripping effect will not influence the results obtained.

The optimal design of flexure hinges in microgripper structures has been investigated in multiple works [34–38]. In the 'hot and cold arm' design, the ratio of the flexure length to the hot arm length is one of the design parameters that influences the maximum deflection achieved at the microgripper arm tips. This length ratio was optimised in the authors' previous work [32] and was found to have a value of 0.23. Moreover, the flexure is generally manufactured to have the same width as the hot arm, which is in turn usually designed with the smallest dimension that the fabrication process is capable of achieving. The flexure is, however, much shorter than the hot arm, thus limiting its temperature from increasing to the same degree as that of the hot arm. These length and width criteria ascertain that the flexure is designed with the correct dimensions to effectively operate as a low-stiffness extension of the cold arm, ensuring proper elastic deflection of the actuator. As for the entire microgripper structure, the thickness of the flexure in this work is dependent on the fabrication process.

The performance of a microgripper is largely dependent on its geometry and dimensional parameters, as well as its material properties. This work focuses on four microgripper design variants that have the same dimensions in the x-y plane (i.e., the length and width parameters), but different structure and air gap thicknesses, as well as different material properties. The fixed length and width parameters of the designed microgrippers are given in Table 1.

Table 1. Fixed geometrical dimensions of the studied microgripper design variants. The dimension annotations are given in Figure 1.

Parameter	Value	Unit
Length of hot arm, L_h	200	µm
Length of cold arm, L_c	154	µm
Length of flexure, L_f	46	µm
Length of connector, L_g	3	µm
Length of gripping arm, L_a	203	µm
Width of hot arm, w_h	3	µm
Width of cold arm, w_c	14	µm
Width of flexure, w_f	3	µm
Width of gripping arm, w_a	6	µm

The differences in structure thickness, air gap thickness, and material type of the four microgripper design variants were achieved by making use of different fabrication technologies: three designs were obtained using the different structural layers offered by the PolyMUMPs^{TM} process, and one design was obtained using the SOIMUMPs^{TM} process. These two fabrication technologies have very different sources of silicon, mainly low pressure chemical vapour deposition (LPCVD) for polycrystalline silicon films for the PolyMUMPs^{TM} process, and mechanically-thinned wafers of single crystal silicon (SCS) for the SOIMUMPs^{TM} process, resulting in different material properties. SOIMUMPs^{TM} utilises a 25 µm thick SCS structure having only one free-standing structural layer, whereas PolyMUMPs^{TM} utilises polysilicon structures with thicknesses in the range of micrometers that are made from two free-standing, independently stackable structural layers. Although the individual sacrificial and structural layer thicknesses are determined by the PolyMUMPs^{TM} fabrication process, three design variants with different thickness dimensions have been fabricated by using the two available structural layers individually as well as by fabricating a stack of both layers, enabling the user to achieve structure thicknesses of 1.5 µm, 2 µm and 3.5 µm with the same process. These two structural layers are both made of polysilicon but are subject to a different amount of diffusion-doping with phosphorus, leading to different values for the electrical resistivity of the two layers. For the same in-plane geometry, the different structural thicknesses result in a wide range of aspect ratio (width:thickness) values, from a value of 30 for the 1.5 µm thick PolyMUMPs^{TM} structure down to a value of 1.8 for the 25 µm thick SOIMUMPs^{TM} structure. Moreover, the different thicknesses of the PolyMUMPs^{TM} structures also result in different air gap separations between the device and the substrate, while no small air gap is present in the SOIMUMPs^{TM} structure.

3. Fabrication Processes

The PolyMUMPs^{TM} and SOIMUMPs^{TM} fabrication technologies are both standard, commercially available Multi-User MEMS Processes (MUMPs®) that are produced by MEMSCAP, Inc.

PolyMUMPs^{TM} [39] is a surface micromachining process that utilises a 20 µm thick (100) SCS handle wafer (known as Substrate), a silicon nitride layer for electrical isolation (known as Nitride), three mechanical layers of LPCVD polysilicon (known as Poly-0, Poly-1 and Poly-2), two sacrificial layers of LPCVD phosphosilicate glass (PSG) (known as Oxide-1 and Oxide-2), and a layer of gold with a thin chromium adhesion layer (known as Metal) that are deposited by electron beam for probing, bonding, or electrical routing. Once the layers are photolitographically patterned and reactive ion etched, the PolyMUMPs[TM] process results in a suspended structure, i.e., a structure that is separated by a small air gap from the surface of the substrate. Because of this small air gap between the device and the substrate, small bumps known as dimples are introduced underneath the polysilicon beam to reduce the contact surface area in case the beam bends towards the substrate. Such dimples thus help to minimise device stiction to the substrate that is driven by surface energy. The Poly-1 and Poly-2 layers can be used individually or stacked on top of each other to create suspended movable structures that are released when the intermediate PSG layers are removed through etching with hydrofluoric (HF) solution, followed by supercritical CO_2 drying.

On the other hand, SOIMUMPs[TM] [40] is a silicon-on-insulator (SOI) micromachining process that utilises a 400 µm thick (100) SCS handle wafer (known as Substrate), a silicon dioxide layer for electrical isolation (known as Oxide), a SCS layer that is mechanically-thinned (known as SOI), and two layers of gold with a thin chromium adhesion layer (known as Pad-Metal and Blanket-Metal) that are deposited by electron beam . The 25 µm thick SOI layer is mechanically released by jointly using an undercut etch with vapour phase HF acid to remove the Oxide layer, and deep reactive ion etching from the backside of the substrate layer (known as Trench), in this way creating an overhanging movable structure, i.e., a structure that is totally exposed to ambient with no small air gap present.

Figure 2 highlights the main dimensional differences resulting from the two fabrication processes. The main differences between structures fabricated with these two technologies are the different thickness of the released silicon layers t, the different air gap thickness t_a or lack thereof, as well as the different structural material. Table 2 summarises the microgripper design variants studied in this work.

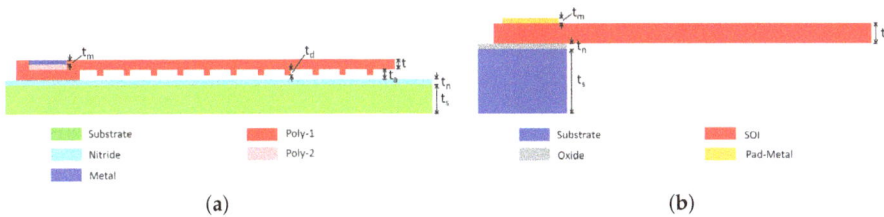

Figure 2. Cross-sectional schematic of the microgripper structure fabricated with the PolyMUMPs[TM] surface micromachining technology (**a**) and the SOIMUMPs[TM] bulk micromachining technology (**b**) (not-to-scale). Note that the PolyMUMPs[TM] process results in a suspended structure (i.e., with a small air gap between the device and the substrate) while the SOIMUMPs[TM] process results in an overhanging structure (i.e., the structure is totally exposed to ambient with no small air gap present).

Table 2. Design variants of the studied microgripper structure. The given nominal thicknesses are in µm and are annotated in Figure 2. The thickness parameters are defined by the respective fabrication process and the mask layers used.

Design Variant #	t	t_a	t_d	t_s	t_n	t_m	Process	Material	Mask Layer
1	1.5	2.75	2	20	0.6	0.5	PolyMUMPs[TM]	Polysilicon	Poly-2
2	2	2	0.75	20	0.6	0.5	PolyMUMPs[TM]	Polysilicon	Poly-1
3	3.5	2	0.75	20	0.6	0.5	PolyMUMPs[TM]	Polysilicon	Poly-1 + Poly-2
4	25	-	-	400	2	0.5	SOIMUMPs[TM]	Single Crystal Silicon	SOI

4. Numerical Models

Numerical models based on the FEA technique were developed in CoventorWare® and used to sequentially compute the steady-state electrothermal and thermomechanical performance of the designed microgrippers. The models of the microgripper design variants fabricated with the PolyMUMPs™ process (Design Variants 1-3) and with the SOIMUMPs™ process (Design Variant 4) are shown in Figures 3 and 4, respectively. A parabolic tetrahedral mesh with an element size of 1 μm was determined by a mesh refinement study to ensure an accurate and cost-effective mesh size. The numerical simulations were performed taking into account nonlinear geometry, while the influence of body forces such as gravity were not considered as they were found to be negligible at the microscale. The arm tip points indicated in Figure 1 were numerically investigated for their *x-*, *y-*, and *z*-displacements.

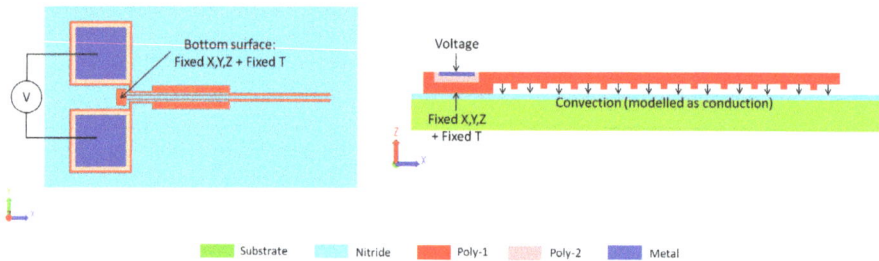

Figure 3. The suspended PolyMUMPs™ electrothermal microgripper as modelled in CoventorWare®. This model represents the mechanical structure for Design Variants 1-3, and demonstrates the boundary conditions applied to the model.

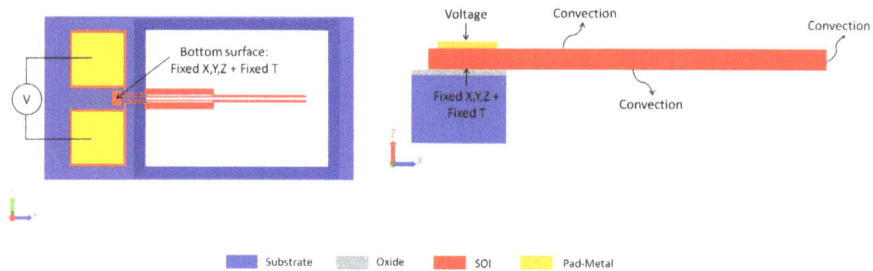

Figure 4. The overhanging SOIMUMPs™ electrothermal microgripper as modelled in CoventorWare®. This model represents the mechanical structure for Design Variant 4, and demonstrates the boundary conditions applied to the model.

The material properties used within the numerical models are given in Table 3 and Figure 5. Isotropic properties are used for polycrystalline silicon. Polysilicon films with a 1 μm thickness or larger are usually composed of relatively columnar grains with an average grain size of 300–600 nm across most of the film thickness and dominant <100> texture. This grain size is strongly influenced by deposition parameters and annealing temperatures. Such a microstructure permits large polysilicon components composed of a large aggregate of grains, such as MEMS devices that are tens to hundreds of micrometers large, to be considered as transversely isotropic and under plane strain, locally and globally [11]. In the case of SCS, while thermal properties can be assumed to be isotropic for any given plane, the elastic and mechanical properties are, however, direction dependent. This variation results from the anisotropic nature of silicon whose crystalline structure exhibits cubic symmetry. The elastic behavior of silicon MEMS structures depends on the orientation of the structure with respect to the crystallographic axes. The (100) plane is the most common silicon wafer orientation used for the microfabrication of MEMS devices and thus the elasticity values of SCS are given in the frame of reference of a standard (100) silicon wafer in this work. Due to its cubic symmetry, it is possible to describe silicon as an orthotropic material and to give its elastic properties in terms of orthotropic material constants as shown in Table 3 [42,43].

Table 3. Material properties extracted from the CoventorWare® Materials Library for the PolyMUMPs™ [39] and SOIMUMPs™ [40] fabrication processes. Unless otherwise specified, these material properties are given at 300 K.

Property	PolyMUMPs™			SOIMUMPs™	
	Poly1	Poly2	Metal	SOI	PadMetal
Density [g/(cm)^3]	2.23	2.23	19.30	2.50	19.30
Young's modulus, E [GPa]	158	158	57	$E_x = E_y = 169$ $E_z = 130$	57
Shear modulus, G [GPa]	-	-	-	$G_{yz} = G_{zx} = 79.6$ $G_{xy} = 50.9$	-
Poisson's ratio, ν	0.22	0.22	0.35	$\nu_{yz} = 0.36, \nu_{zx} = 0.29$ $\nu_{xy} = 0.064$	0.35
Thermal expansion coefficient, α [µm/mK]	2.80	2.80	Refer to Figure 5	2.50	Refer to Figure 5
Specific heat capacity, c [J/kgK]	712	712	128.7	712	128.7
Thermal conductivity, k [W/mK]	32	32	297	148	297
Electrical resistivity [µ·Ω·m]	20	30	3.12×10^{-2}	500	2.86×10^{-2}

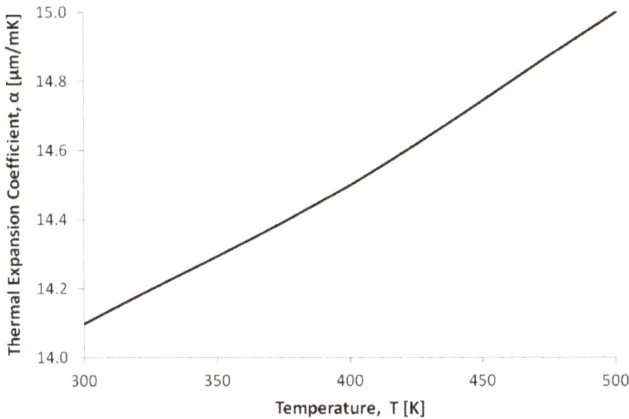

Figure 5. Variation of the coefficient of thermal expansion, α, with temperature for the Metal and PadMetal Layers in the PolyMUMPs™ [39] and SOIMUMPs™ [40] processes, respectively.

The solution of the electrothermal and thermomechanical analyses also requires a number of boundary conditions to be applied within the models to define the electrical, thermal and mechanical constraints imposed on the system. The boundary conditions implemented within these models are indicated in Figures 3 and 4, and included applying the actuation voltage to the top surface of the metallised probe pads, setting an initial temperature as a volumetric boundary condition to all the model elements, fixing the temperature of the bottom surface of the three anchored pads to room temperature throughout the analyses since these pads are anchored to the substrate, which acts as a heat sink, and restricting the motion of the bottom surface of the three anchored pads in all directions throughout the analyses. All the above-mentioned boundary conditions were identically applied to the four microgripper design variants.

An additional boundary condition was then separately applied to the PolyMUMPs^{TM} structures (Design Variants 1-3) and the SOIMUMPs^{TM} structure (Design Variant 4) to simulate convective heat transfer. Due to the very small air gap, t_a, between the microgripper and the substrate resulting from the PolyMUMPs^{TM} fabrication technology (Figure 2a), convection from the bottom surface of the microgripper to the substrate becomes significant, and other heat losses by natural convection from the other beam surfaces can justifiably be assumed to be comparatively negligible. Moreover, the small air gap permits the convective heat transfer coefficient of the bottom surface to be modelled by conduction through the air gap to the substrate [44]. This approach was thus applied to simulate heat losses in the numerical models for Design Variants 1-3. On the other hand, due to the absence of a small air gap beneath the overhanging SOIMUMPs^{TM} structure (Figure 2b), there is no dominant convective heat loss from the bottom surface compared to the other beam surfaces and thus, in this case, a convective heat transfer coefficient was equally applied to all the beam surfaces in the numerical model for Design Variant 4 to simulate heat losses by natural convection in air. Moreover, heat losses by radiation can be reasonably ignored for thermal actuators operated at low operational power [44,45] and were thus excluded from all the models in this work.

5. Experimental Setup

The experimental testing reported in this work was carried out on the microgripper structures fabricated with PolyMUMPs^{TM} (Design Variants 1-3). Different laboratory procedures are in place to reliably test RBCs under atmospheric conditions [46], thus justifying actuation characterisation in air in this study. The probe pads of the released fabricated structures were ultrasonically wire-bonded to the contact pads of an integrated circuit (IC) package to facilitate testing. Voltage probes, each with an integrated positioning system in the x-, y- and z-directions, were then used to electrically achieve a connection between the IC package and the power supply. Each microgripper structure based on Design Variants 1-3 was actuated by applying a potential in increments of 0.5 up to 3 V on each arm, in the meantime also taking note of the current through the structure. The thermal time constant of the studied microgripper design variants is of the order of milliseconds, and thus an interval of a few seconds between each voltage step ensured that steady-state conditions were obtained before applying the new voltage.

The relative opening displacement of each microgripper arm tip was measured at atmospheric pressure on a vibration isolation Cascade Microtech Summit 11,000/12,000 B-series probe station with an embedded optical microscope-based vision system. A Mitutoyo's Metallurgical Plan Apochromatic (M Plan APO) 10× objective lens, in conjunction with the 10× magnification of the eyepiece, was used for the experimental characterisation. The features of such an objective lens include infinity corrections, bright field observation, long working distance and corrections of chromatic and spherical aberrations. A tracking software program was used to automatically measure the relative opening stroke of each arm, from which the absolute total gap opening could then be calculated by taking into account the 5 μm initial gap between the arm tips in the closed (i.e., not actuated) position. The width of the gripping arm (6 μm) was taken as a reference for calibration. The experimental setup is shown in Figure 6.

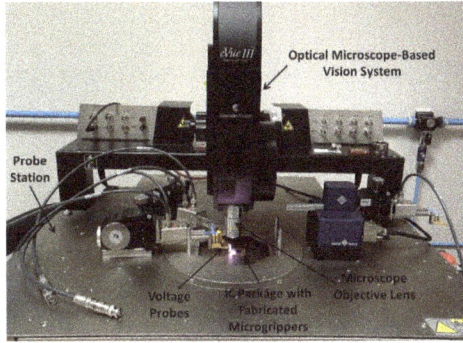

Figure 6. Experimental setup used to test the microgripper design variants fabricated with PolyMUMPs™. The microgripper structures are mounted on an integrated circuit package and actuated with voltage probes on the Cascade Microtech probe station [25].

6. Results and Discussion

The four microgripper design variants that differ in terms of the device thickness, the air gap thickness between the device and the substrate or the lack thereof, and the silicon type (polysilicon versus SCS) have been studied for their electrothermomechanical performance. This has been conducted analytically, numerically and experimentally for Design Variants 1-3, and analytically and numerically for Design Variant 4.

6.1. Thermal Analyses

The temperature distribution developed within the microgripper is one of the important factors to be considered in a microgripper design. The maximum temperature on the hot arm determines the microgripper's performance, but, at the same time, this temperature needs to be kept within the material limits to avoid the onset of material damage. Additionally, for specific applications in the biomedical field, the temperature at the microgripper arm tips needs to be kept as minimum as possible to ensure the manipulation of living cells and tissues without inducing any damage.

Thermal characterisation of the actuated microgripper structures was performed analytically and numerically in this work. Figure 7 compares the analytical results for the temperature distribution along the different actuator components for Design Variants 1-4. The temperature profiles along the hot arm $T_h(x)$, cold arm $T_c(x)$, and flexure $T_f(x)$ for these structures were calculated using Equations (1)–(3), respectively, as derived in [25]:

$$T_h(x) = T_H + C_1 e^{m_h x} + C_2 e^{-m_h x}, \tag{1}$$

$$T_c(x) = T_C + C_3 e^{m_c x} + C_4 e^{-m_c x}, \tag{2}$$

$$T_f(x) = T_F + C_5 e^{m_f x} + C_6 e^{-m_f x}, \tag{3}$$

where m_h, m_c and m_f are to be calculated for the hot arm, cold arm and flexure, respectively, using:

$$m = \sqrt{\frac{S}{k_p t R_t}}, \tag{4}$$

where S is the shape factor that accounts for the impact of the shape of the hot arm/cold arm/flexure on heat conduction to the substrate, k_p is the thermal conductivity of polysilicon, t is the device thickness and R_t is the thermal resistance between the polysilicon beam and the substrate. The constants C_1 to C_6

are calculated based upon the boundary conditions applied to the actuator structure as detailed in [25]. In the case of Design Variant 4, a very large air gap thickness of 400 µm was used in the analytical model, in this way rendering heat conduction from the bottom surface of the device through the large air gap negligible as is the case for an overhanging structure.

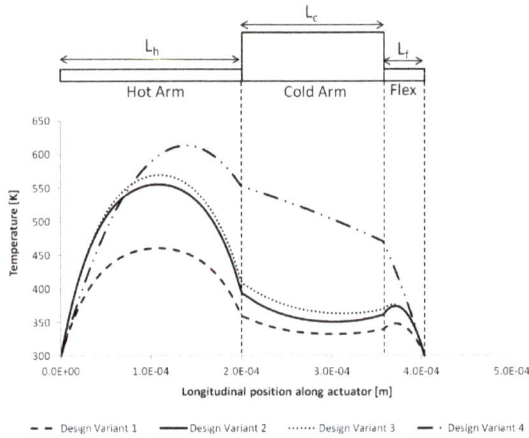

Figure 7. Comparison of the analytical results for the temperature profile along the hot arm, cold arm and flexure components of Design Variants 1-3 at 3 V, and of Design Variant 4 at 14 V. The diagram above the graph illustrates an unfolded thermal actuator without the extending microgripper arms.

The temperature distribution results obtained from the developed numerical models of the PolyMUMPs™ structures (Design Variants 1-3) and the SOIMUMPs™ structure (Design Variant 4) in CoventorWare® are given in Figures 8 and 9, respectively.

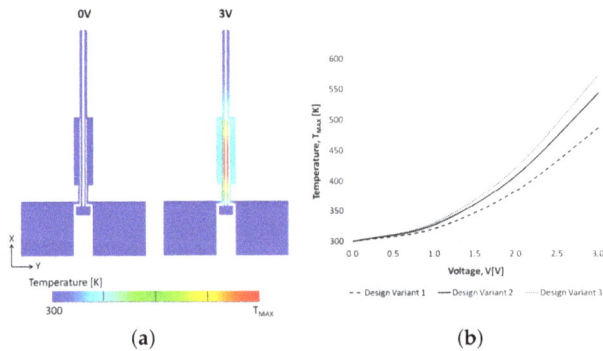

(a) (b)

Figure 8. Simulated steady-state temperature distribution on the polysilicon microgripper structures fabricated with PolyMUMPs™ (Design Variants 1-3). The simulated plots in (**a**) show the temperature distribution with and without an applied potential for Design Variants 1-3. The maximum temperature of the actuated microgripper structure is located on the hot arm and is given by T_{MAX}, which is then quantified with applied potential in (**b**) for each specific design variant.

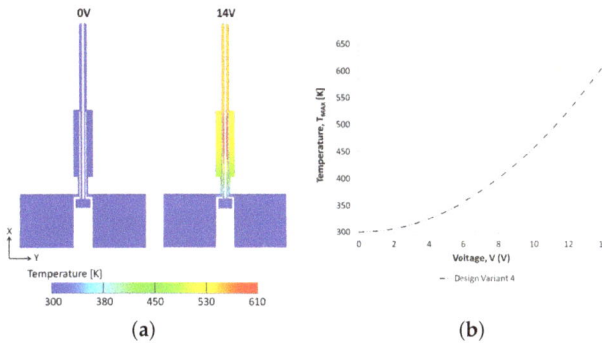

Figure 9. Simulated steady-state temperature distribution on the single crystal silicon microgripper structure fabricated with SOIMUMPs^{TM} (Design Variant 4): simulated temperature plots showing the distribution with and without an applied potential (**a**), and variation of the maximum temperature, T_{MAX}, with applied potential (**b**). The maximum temperature, T_{MAX}, of the actuated microgripper structure at 14 V is 610 K and is located on the hot arm as shown in (**a**).

It can be observed that, for each microgripper design variant, the maximum temperature is always located on the hot arm, independent of the structure thickness and the fabrication technology used (Figures 7, 8a and 9a). This is an inherent characteristic of the 'hot and cold arm' actuator design. In this design, the temperature difference between the hot and cold arms plays an important role in the achieved end-effector lateral displacement, with the latter increasing as the stated temperature difference increases for the same applied voltage. One way to maximise this temperature difference is to design the hot arm in such a way that its cross-sectional area is as small as possible considering the capability of the fabrication process. In the case of PolyMUMPs^{TM} and SOIMUMPs^{TM} fabrication technologies, the recommended minimum feature width is 3 µm [39,40], which is, in fact, the value used for the width of the hot arm for the four microgripper design variants in this study (Table 1).

When considering the microgrippers fabricated with the PolyMUMPs^{TM} technology (Design Variants 1-3), it can be observed that the most distinguished difference is between the temperature distribution of Design Variant 1, and that of Design Variants 2 and 3. The structure thickness, the air gap thickness and the intrinsic nature of the materials used (mainly in terms of the electrical resistivity property) all play a major role in defining the obtained temperature distribution on the microgripper structure. The maximum temperature developed on the hot arm increases with a thicker structure, a larger air gap thickness due to less heat lost by conduction through the air gap, and a smaller value for the electrical resistivity material property. When compared to Design Variants 2 and 3, Design Variant 1 has the smallest structure thickness of 1.5 µm, the largest air gap thickness of 2.75 µm and the highest value for electrical resistivity (Table 3). The effects of the higher electrical resistivity value and the smaller structure thickness outweigh the effect of the larger air gap thickness, resulting in an overall reduction in the maximum temperature on the hot arm. The temperature distribution for Design Variants 2 and 3 is similar. These two microgripper design variants have the same air gap thickness (2 µm), which affects the amount of heat lost by conduction to the substrate. The different thicknesses (2 µm for Design Variant 2 and 3.5 µm for Design Variant 3) and electrical resistivities (20 µΩm for Design Variant 2 and 23.33 µΩm for Design Variant 3) of the two structures almost balance out each other, resulting in a similar temperature distribution.

On comparing the temperature distribution of a surface micromachined structure (Design Variants 1-3) as given in Figures 7 and 8, to that of a bulk micromachined structure (Design Variant 4) as given in Figures 7 and 9, it can be observed that, despite the larger structure thickness and lack of a small air gap within the SOIMUMPs^{TM} structure, the larger electrical resistivity value of SCS when compared to that of polysilicon (Table 3) results in greater power required to achieve a quantitatively similar temperature distribution to that obtained for the PolyMUMPs™ structures. It can be also observed from Figure 7 that the temperatures on the cold arm and the flexure are relatively high for Design Variant 4 when compared to those for Design Variants 1-3. This results in Design Variant 4 having the smallest temperature difference between the hot and cold arms, and, as stated previously, the larger this temperature difference, the larger will be the achieved end-effector lateral displacement. This small temperature difference will in fact cause the tips of Design Variant 4 to undergo the smallest opening stroke as will be shown in Section 6.2. Another interesting observation is that while the arm tips of the actuated PolyMUMPs™ structures remain at room temperature (Figure 8a), this is not the case for the SOIMUMPs™ structure. It is shown in Figure 9a that the arm tips of the actuated SOIMUMPs™ microgripper undergo a rise in temperature due to the higher thermal conductivity of SCS when compared to that of polysilicon (Table 3). The lower thermal conductivity value of polycrystalline silicon is a result of the grain boundary scattering of phonons that governs the thermal resistance [47]. It must be ensured that the temperature rise at the arm tips is limited as much as possible to prevent damage to the studied RBCs.

6.2. Structural Analyses

A horizontal MEMS microgripper should always be designed such that it maximises the in-plane displacement and minimises the out-of-plane deflection at the microgripper am tips. The steady-state lateral displacement at the gripping arm tip, u_{tip}, was calculated as a function of applied potential using Equation (5) as derived in [25]:

$$u_{tip} = \frac{L_h^2}{6EI_h} \left(X_1 L_h - 3X_3 \right) + \frac{L_a L_h}{2EI_h} \left(L_h X_1 - 2X_3 \right), \tag{5}$$

where L_h is the length of the hot arm, E is the Young's modulus of polysilicon, I_h is the moment of inertia for the hot arm, and X_1, X_2 and X_3 are the chosen redundant loads acting on the actuator structure. The force method [48] was applied to analyse the bending moment of the actuator due to these three redundant loads, which were calculated by solving the following set of simultaneous equations as detailed in [25]:

$$\begin{bmatrix} f_{11} & f_{12} & f_{13} \\ f_{21} & f_{22} & f_{23} \\ f_{31} & f_{32} & f_{33} \end{bmatrix} \begin{bmatrix} X_1 \\ X_2 \\ X_3 \end{bmatrix} = \begin{bmatrix} 0 \\ \Delta L_h - \Delta L_c - \Delta L_f \\ 0 \end{bmatrix}, \tag{6}$$

where the terms f_{ij} represent flexibility coefficients that define the deflection at i due to a unit load at j. The flexibility coefficients are determined using Equation (7) where m_i and m_j represent the bending moments due to the respective three unit redundants X_1, X_2 and X_3:

$$f_{ij} = \int_L \frac{m_i m_j}{EI} dx. \tag{7}$$

The relative opening stroke of each arm tip, calculated using Equation (5), is compared for Design Variants 1-4 in Figure 10. The steady-state in-plane and out-of-plane displacements are also obtained numerically with CoventorWare® for Design Variants 1-4 (Figures 11–13), and good agreement is observed for the obtained in-plane displacement between the analytical and numerical results. The current and power dissipated across each microgripper arm of Design Variants 1-4 with applied voltage are given in Figure 14.

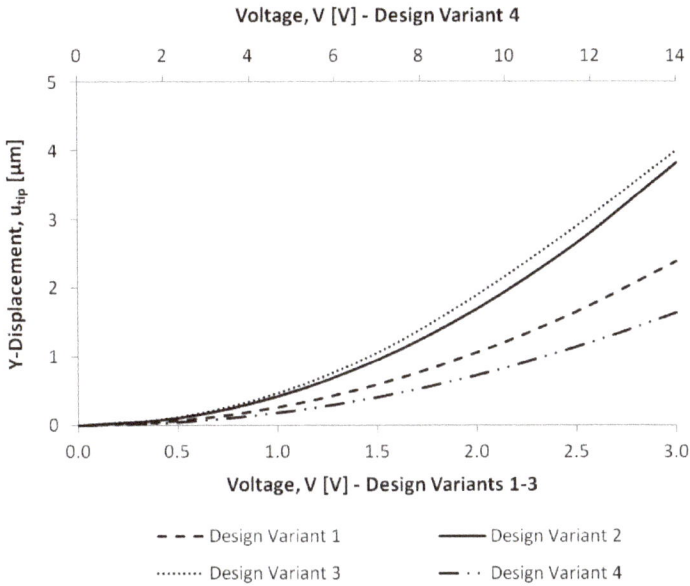

Figure 10. Analytical results for the variation of the relative *y*-displacement of the arm tip, u_{tip}, with applied voltage for Design Variants 1-4.

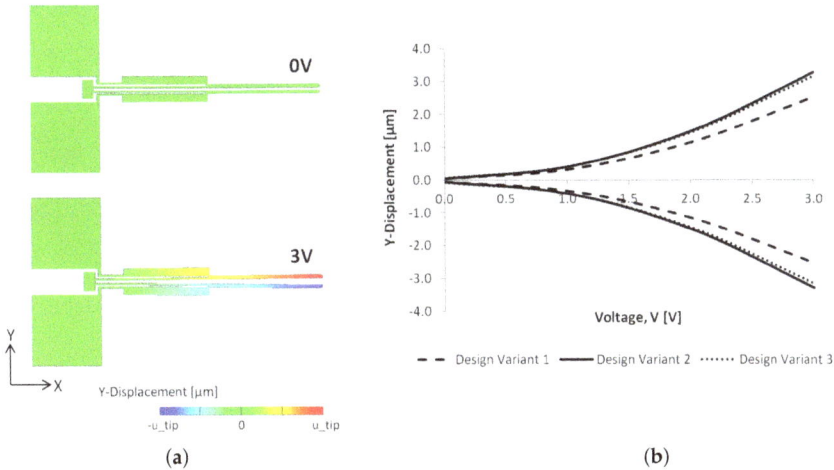

Figure 11. Simulated steady-state lateral displacement distribution on the polysilicon microgripper structures fabricated with PolyMUMPs[TM] (Design Variants 1-3). The simulated plots in (**a**) show the *y*-displacement distribution with and without an applied potential for Design Variants 1-3. The absolute gap opening in the closed position (i.e., when not actuated), u_o, is 5 μm with this absolute gap increasing to $u_o + 2u_{tip}$ when actuated. The relative opening stroke of each microgripper arm tip is then quantified with applied potential in (**b**) for each specific design variant.

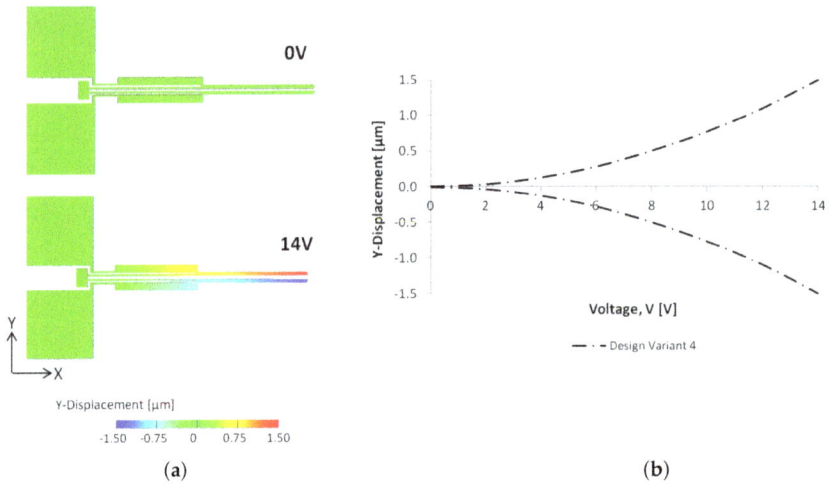

Figure 12. Simulated steady-state lateral displacement distribution on the single crystal silicon microgripper structure fabricated with SOIMUMPs™ (Design 4): simulated *y*-displacement plots showing the distribution with and without an applied potential (**a**); and variation of the relative opening stroke of each microgripper arm tip with applied voltage (**b**). The absolute gap opening in the closed position is 5 μm with this absolute gap increasing to 8 μm at 14 V.

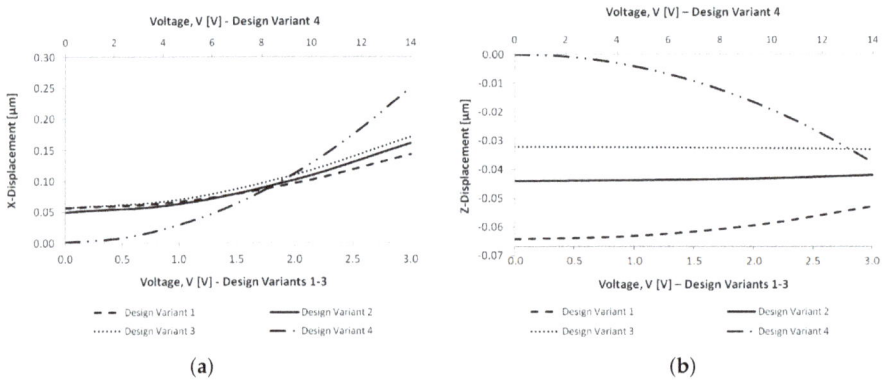

Figure 13. Simulated steady-state displacement profiles of Design Variants 1-4 with applied potential: *x*-displacement (forward displacement) of the microgripper arm tips (**a**); and *z*-displacement (out-of-plane displacement) of the microgripper arm tips (**b**).

On comparing the relative arm lateral displacements of Design Variants 1-4 (Figures 10–12), it is evident that Design Variant 4 undergoes the smallest opening stroke when actuated. In the case of Design Variant 4, the total gap opening is analytically and numerically predicted to be around 8 μm at an applied potential of 14 V, which is just in line with the design requirement of gripping a RBC. On the other hand, Design Variants 1-3 exceed the basic requirement of 8 μm for the absolute gap opening between the arm tips. Despite this, it can be observed that SOIMUMPs™ structures require much greater power than PolyMUMPs™ structures (Figure 14) to achieve the intended motion. Another important observation is the difference in out-of-plane displacement (*z*-displacement) at 0 V for Design Variants 1-4 (Figure 13b). This difference can be attributed to the inherent greater thickness and lower aspect ratio (width:thickness) fabrication of SOIMUMPs™ structures when compared to PolyMUMPs™ structures. The aspect ratio for PolyMUMPs™ structures depends on the layers used, but it is always much larger than the aspect ratio for SOIMUMPs™ structures. The greater moment of inertia of structures with low aspect ratio, such as those fabricated with SOIMUMPs™, minimises any unwanted out-of-plane displacement of the actuator arms as well as offers better resistance to the effects of in-plane residual stress, making such structures more suitable for in-plane directed thermal actuators [49]. It can also be observed from Figure 13b that Design Variant 3, which is composed of the stacked Poly-1 and Poly-2 layers, has the least out-of-plane displacement when compared to Design Variants 1 and 2, which confirms that lower aspect ratio structures minimise out-of-plane bending occurring as a result of residual stresses inherent in the fabrication process. Minimising the out-of-plane path deviation of the microgripper arms is an important requirement for the successful gripping of RBCs to ensure that the arm tips remain on the same plane as the RBC so that the latter can be successfully gripped during actuation.

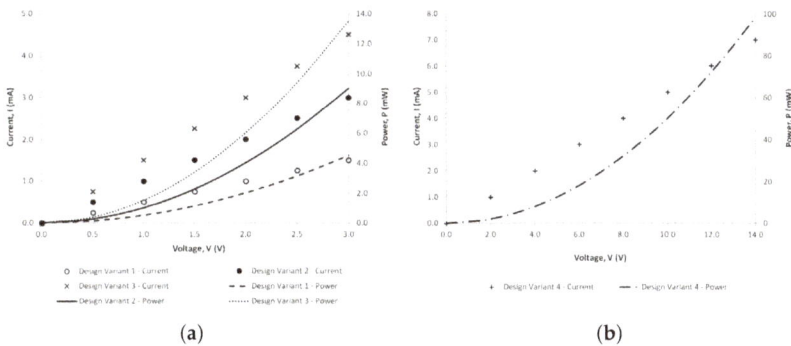

Figure 14. Numerical variation of the current and power dissipated with voltage across each arm of the microgripper structure for Design Variants 1-3 (**a**), and for Design Variant 4 (**b**).

Figure 15 shows the investigation of the developed maximum stresses within the 'hot and cold arm' microgripper structure. It can be observed that the maximum stresses are mainly located where the arms join the anchor pads, within the flexure component and at the connectors joining the hot and cold arms. The maximum Von Mises stress for Design Variants 1-4 ranges between 40 MPa and 60 MPa at full actuation voltage (3 V for Design Variants 1-3, 14 V for Design Variant 4). The ultimate tensile strength for polysilicon and SCS has been reported in literature as 2.9 GPa [50] and 2 GPa [51], respectively. This indicates that the simulated maximum stress values are by far within the limits of the allowable stress, concluding that maximum stresses and strains during the opening motion are not critical.

The analytical and numerical models of the PolyMUMPs™ structures (Design Variants 1-3) were validated through optical microscopy studies carried out under atmospheric pressure using the setup described in Section 5. Figure 16 shows the optical microscope images of Design 2 in the closed and actuated positions. The absolute gap opening between the arm tips when the microgripper is not actuated is 5 μm. The simulation and experimental measurement results of the absolute gap opening between the microgripper arm tips are compared for Design Variants 1-4 in Figure 17. Figure 17 demonstrates that the mean absolute gap openings obtained experimentally at 3 V are 9.5 μm for Design 1, 12.5 μm for Design 2 and 12 μm for Design 3, and that the results predicted by the numerical models are in good agreement with the experimental results for Design Variants 1-3. The measured gap opening values confirm that Design Variants 1-3 are all suitable for the desired function of securing a RBC with an approximate 8 μm diameter.

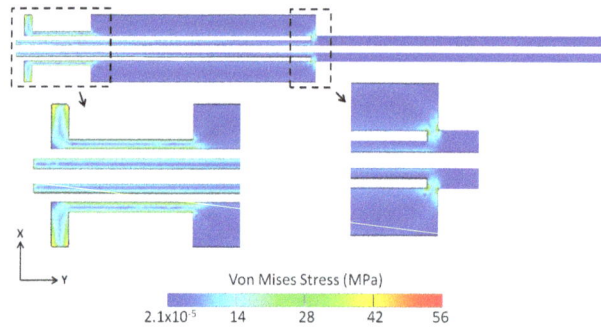

Figure 15. Von Mises stresses developed within the microgripper structure for Design Variant 1 at 3 V with inset images showing detailed views of the stresses where the arms join the anchor pads, within the flexure component and at the connectors joining the hot and cold arms.

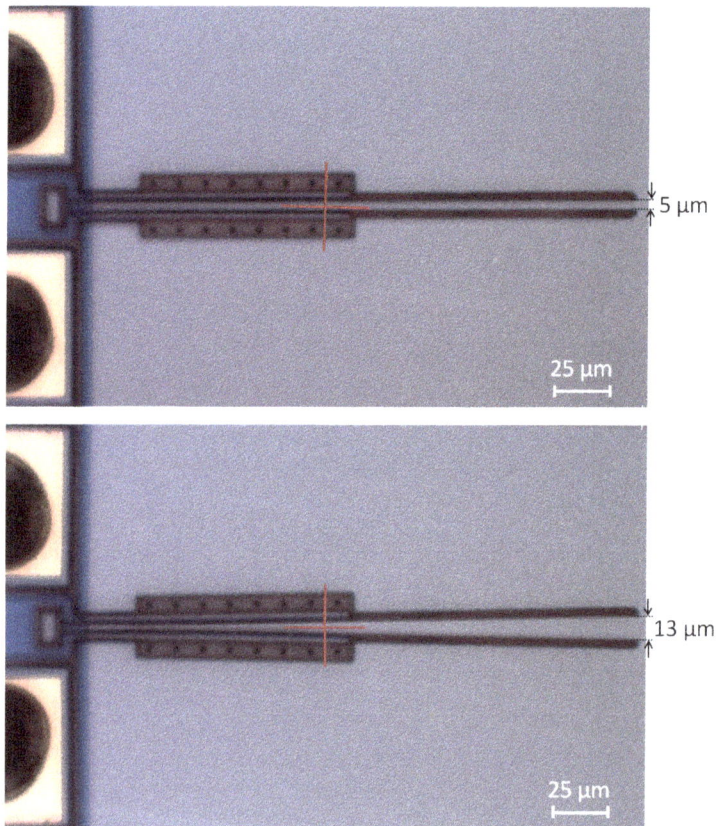

Figure 16. Images based on optical microscopy of one of the fabricated microgripper structures (Design 2) when not actuated (**top**), and when each arm is electrothermally actuated with 3 V (**bottom**). The pixel resolution of the vision system photographs is 1023 pixels (width) × 767 pixels (height), and the marked values give the absolute gap openings between the microgripper arm tips.

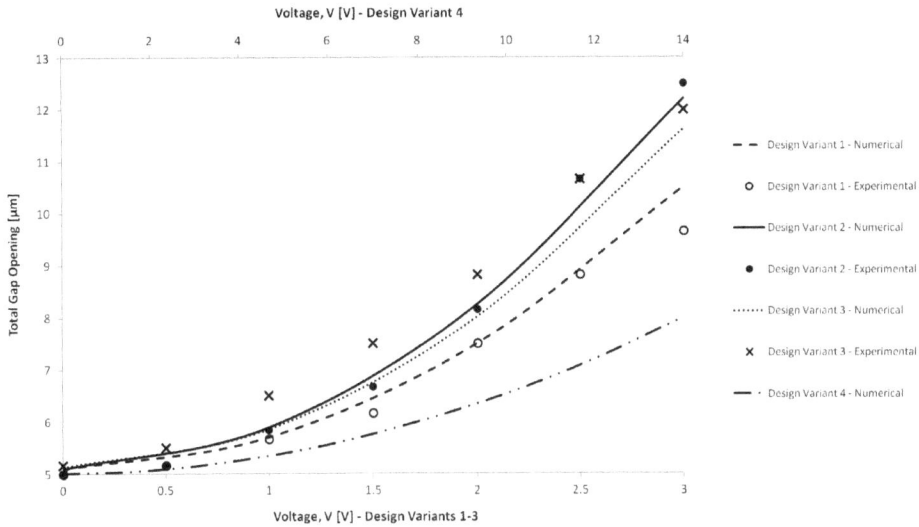

Figure 17. Comparison of the experimental and numerical results for the absolute gap opening between the arm tips of the studied microgripper design variants with applied voltage. Experimental (mean values) and numerical results are presented for Design Variants 1-3 while numerical results are shown for Design Variant 4. All design variants start from an absolute gap opening of 5 μm, which is the gap opening in the closed position.

7. Conclusions

This paper has presented four design variants of a horizontal electrothermally actuated MEMS microgripper that were developed for the micromanipulation and deformability study of RBCs using two commercially available fabrication technologies. Three microgripper design variants were achieved using the different stackable structural layers offered by the PolyMUMPs^{TM} process, resulting in polysilicon structure thicknesses of 1.5 μm, 2 μm and 3.5 μm. The other microgripper design variant was achieved using the SOIMUMPs^{TM} process resulting in a SCS structure with a 25 μm thickness. The studied microgripper design variants have the same in-plane geometry and their main differences are due to the mechanical structure thickness, the air gap thickness or lack thereof, and the silicon type (polysilicon versus SCS), all of which are a result of the inherent characteristics of the specific fabrication technology used.

The electrothermomechanical performance, specifically the power dissipation, the temperature distribution, the lateral opening stroke and the out-of-plane deflection at the microgripper arm tips, were examined and compared for the microgripper design variants. Analytical calculations and FEA simulations with the commercial software package CoventorWare® were performed on the four microgripper design variants to investigate the steady-state response of the microgripper structures under electrothermal actuation. It could be observed that greater power is required for the SOIMUMPs^{TM} structure to achieve a quantitatively similar temperature distribution and tip displacement to that obtained from the PolyMUMPs^{TM} structures. Another interesting observation was that the microgripper arm tips of the SOIMUMPs^{TM} structure, unlike PolyMUMPs^{TM} structures, undergo a rise in temperature, making such a design less suitable for the safe manipulation of biological cells. However, SOIMUMPs^{TM} structures have lower aspect ratios when compared to PolyMUMPs^{TM} structures. This means that the greater moment of inertia of SOIMUMPs^{TM} structures results in substantial out-of-plane stiffness, thus minimising any undesired out-of-plane path deviation and ensuring a more stable in-plane motion. The maximum stresses developed within the microgripper structure were also investigated for all the design variants and they were found to be well within the allowable material limits, confirming that the microgripper design variants are able to perform a stable opening and closing motion.

The three microgripper design variants based on the PolyMUMPs™ process were fabricated, and were subject to optical microscopy studies under atmospheric pressure. The tip displacement results from the actuation testing were found to comply with the analytical and numerical predictions, and all the design variants were also found to be in line with the design requirement to secure and characterise a RBC. However, it can be concluded that Design 3, which consists of two polysilicon stackable layers, resulting in a 3.5 µm thick PolyMUMPs™ structure, best fulfills the desired function. This design experimentally achieved a sufficient mean gap opening of 12 µm at 3 V, while, at the same time, it was numerically found to limit the out-of-plane motion at the arm tips and to ensure room temperature conditions at the cell gripping zone.

Author Contributions: Conceptualisation, M.C., I.G., B.M., P.M. and N.S.; Formal analysis, M.C.; Funding acquisition, M.C., I.G., B.M., P.M. and N.S.; Investigation, M.C.; Methodology, M.C.; Project administration, M.C., I.G., B.M., P.M. and N.S.; Resources, M.C.; Supervision, I.G., B.M., P.M. and N.S.; Validation, M.C.; Visualisation, M.C.; Writing—Original draft, M.C.; Writing—Review and editing, M.C., I.G., B.M., P.M. and N.S.

Funding: The research work disclosed in this publication is funded by the Reach High Scholars Programme— Post-Doctoral Grants. The grant is part-financed by the European Union, Operational Programme II—Cohesion Policy 2014–2020 Investing in human capital to create more opportunities and promote the wellbeing of society— European Social Fund (ESF): grant number 254/15/05.

Acknowledgments: The technical help of Barnaby Portelli in the experimental activities performed at the Microelectronics Lab is acknowledged.

Conflicts of Interest: The authors declare no conflict of interest.

References

1. Zhang, R.; Chu, J.; Wang, H.; Chen, Z. A multipurpose electrothermal microgripper for biological micro-manipulation. *Microsyst. Technol.* **2013**, *19*, 89–97. [CrossRef]
2. Verotti, M.; Dochshanov, A.; Belfiore, N.P. Compliance synthesis of CSFH MEMS-based microgrippers. *J. Mech. Des. Trans. ASME* **2017**, *139*, 022301. [CrossRef]
3. Di Giamberardino, P.; Bagolini, A.; Bellutti, P.; Rudas, I.J.; Verotti, M.; Botta, F.; Belfiore, N.P. New MEMS tweezers for the viscoelastic characterization of soft materials at the microscale. *Micromachines* **2017**, *9*, 15. [CrossRef]
4. Kim, K.; Liu, X.; Zhang, Y.; Sun, Y. Nanonewton force-controlled manipulation of biological cells using a monolithic MEMS microgripper with two-axis force feedback. *J. Micromech. Microeng.* **2008**, *18*, 055013. [CrossRef]
5. Kim, K.; Liu, X.; Zhang, Y.; Cheng, J.; Yu, W.X.; Sun, Y. Elastic and viscoelastic characterization of microcapsules for drug delivery using a force-feedback MEMS microgripper. *Biomed. Microdevices* **2009**, *11*, 421–427. [CrossRef] [PubMed]
6. Solano, B.; Wood, D. Design and testing of a polymeric microgripper for cell manipulation. *Microelectron. Eng.* **2007**, *84*, 1219–1222. [CrossRef]
7. Iamoni, S.; Somà, A. Design of an electro-thermally actuated cell microgripper. *Microsyst. Technol.* **2014**, *20*, 869–877. [CrossRef]
8. Cecchi, R.; Verotti, M.; Capata, R.; Dochshanov, A.; Broggiato, G.B.; Crescenzi, R.; Balucani, M.; Natali, S.; Razzano, G.; Lucchese, F.; et al. Development of micro-grippers for tissue and cell manipulation with direct morphological comparison. *Micromachines* **2015**, *6*, 1710–1728. [CrossRef]
9. Potrich, C.; Lunelli, L.; Bagolini, A.; Bellutti, P.; Pederzolli, C.; Verotti, M.; Belfiore, N.P. Innovative silicon microgrippers for biomedical applications: Design, mechanical simulation and evaluation of protein fouling. *Actuators* **2018**, *7*, 12. [CrossRef]

10. Ivanova, K.; Ivanov, T.; Badar, A.; Volland, B.E.; Rangelow, I.W.; Andrijasevic, D.; Sümecz, F.; Fischer, S.; Spitzbart, M.; Brenner, W.; et al. Thermally driven microgripper as a tool for micro assembly. *Microelectron. Eng.* **2006**, *83*, 1393–1395. [CrossRef]

11. Zhang, Y.; Chen, B.K.; Liu, X.; Sun, Y. Autonomous robotic pick-and-place of microobjects. *IEEE Trans. Robot.* **2010**, *26*, 200–207. [CrossRef]

12. Hamedi, M.; Vismeh, M.; Salimi, P. Design, analysis and fabrication of silicon microfixture with electrothermal microclamp cell. *Microelectron. Eng.* **2013**, *111*, 160–165. [CrossRef]

13. Verotti, M.; Dochshanov, A.; Belfiore, N.P. A comprehensive survey on microgrippers design: Mechanical structure. *J. Mech. Des.* **2017**, *139*, 060801. [CrossRef]

14. Jia, Y.; Xu, Q. MEMS microgripper actuators and sensors: The state-of-the-art survey. *Recent Patents Mech. Eng.* **2013**, *6*, 132–142. [CrossRef]

15. Yang, S.; Xu, Q. A review on actuation and sensing techniques for MEMS-based microgrippers. *J. Micro-Bio Robot.* **2017**, *13*, 1–14. [CrossRef]

16. Nguyen, N.-T.; Ho, S.-S.; Low, C.L. A polymeric microgripper with integrated thermal actuators. *J. Microelectromech. Syst.* **2004**, *14*, 969–974. [CrossRef]

17. Somà, A.; Iamoni, S.; Voicu, R.; Müller, R.; Al-Zandi, M.H.M.; Wang, C. Design and experimental testing of an electro-thermal microgripper for cell manipulation. *Microsyst. Technol.* **2018**, *24*, 1053–1060. [CrossRef]

18. Solano, B.; Merrell, J.; Gallant, A.; Wood, D. Modelling and experimental verification of heat dissipation mechanisms in an su-8 electrothermal microgripper. *Microelectron. Eng.* **2014**, *124*, 90–93. [CrossRef]

19. Al-Zandi, M.H.M.; Wang, C.; Voicu, R.; Müller, R. Measurement and characterisation of displacement and temperature of polymer based electrothermal microgrippers. *Microsyst. Technol.* **2018**, *24*, 379–387. [CrossRef]

20. Dochshanov, A.; Verotti, M.; Belfiore, N.P. A comprehensive survey on microgrippers design: Operational strategy. *J. Mech. Des.* **2017**, *139*, 070801. [CrossRef]

21. Kim, C.-J.; Pisano, A.P.; Muller, R.S. Silicon-processed overhanging microgripper. *J. Microelectromech. Syst.* **1992**, *1*, 31–36. [CrossRef]

22. Bagolini, A.; Ronchin, S.; Bellutti, P.; Chistè, M.; Verotti, M.; Belfiore, N.P. Fabrication of novel MEMS microgrippers by deep reactive ion etching with metal hard mask. *J. Microelectromech. Syst.* **2017**, *26*, 7920329. [CrossRef]

23. Wierzbicki, R.; Houston, K.; Heerlein, H.; Barth, W.; Debski, T.; Eisinberg, A.; Menciassi, A.; Carrozza, M.C.; Dario, P. Design and fabrication of an electrostatically driven microgripper for blood vessel manipulation. *Microelectron. Eng.* **2006**, *83*, 1651–1654. [CrossRef]

24. Xu, Q. Design, Fabrication, and testing of an MEMS microgripper with dual-axis force sensor. *IEEE Sens. J.* **2015**, *15*, 7150331. [CrossRef]

25. Cauchi, M.; Grech, I.; Mallia, B.; Mollicone, P.; Sammut, N. Analytical, Numerical and Experimental Study of a Horizontal Electrothermal MEMS Microgripper for the Deformability Characterisation of Human Red Blood Cells. *Micromachines* **2018**, *9*, 108. [CrossRef]

26. Feng, Y.-Y.; Chen, S.-J.; Hsieh, P.-H.; Chu, W.-T. Fabrication of an electro-thermal micro-gripper with elliptical cross-sections using silver-nickel composite ink. *Sensors Actuators A Phys.* **2016**, *245*, 106–112. [CrossRef]

27. Chronis, N.; Lee, L.P. Electrothermally activated SU-8 microgripper for single cell manipulation in solution. *J. Microelectromech. Syst.* **2005**, *14*, 857–863. [CrossRef]

28. Hannon, B.; Ruth, M. Malaria and Sickle Cell Anemia. In *Dynamic Modeling of Diseases and Pests*; Springer: New York, NY, USA, 2009; pp. 63–81.

29. Tomaiuolo, G. Biomechanical properties of red blood cells in health and disease towards microfluidics. *Biomicrofluidics* **2014**, *8*, 051501. [CrossRef] [PubMed]

30. Jeongho, K.; HoYoon, L.; Sehyun, S. Advances in the measurement of red blood cell deformability: A brief review. *J. Cell. Biotechnol.* **2015**, *1*, 63–79.

31. Alizadehrad, D.; Imai, Y.; Nakaaki, K.; Ishikawa, T.; Yamaguchi, T. Quantification of red blood cell deformation at high-hematocrit blood flow in microvessels. *J. Biomech.* **2012**, *45*, 2684–2689. [CrossRef] [PubMed]

32. Cauchi, M.; Mollicone, P.; Grech, I.; Mallia, B.; Sammut, N. Design and analysis of a MEMS-based electrothermal microgripper. In Proceedings of the International CAE Conference 2016, Parma, Italy, 17–18 October 2016.

33. Keoschkerjan, R.; Wurmus, H. A novel microgripper with parallel movement of gripping arms. In Proceedings of the Actuator 2002: 8th International Conference on New Actuators, Bremen, Germany, 10–12 June 2002; pp. 321–324.

34. Lobontiu, N. *Compliant Mechanisms: Design of Flexure Hinges*; CRC Press LLC: Boca Raton, FL, USA, 2003.

35. Meng, Q.; Li, Y.; Xu, J. New empirical stiffness equations for corner-filleted flexure hinges. *J. Mech. Sci.* **2013**, *4*, 345–356. [CrossRef]

36. Figueredo, S. Heat transfer strategies for temperature sensitive components in vacuum environments. In Proceedings of the 16th International Conference of the European Society for Precision Engineering and Nanotechnology, Nottingham, UK, 30 May–3 June 2016.

37. Linß, S.; Schorr, P.; Zentner, L. General design equations for the rotational stiffness, maximal angular deflection and rotational precision of various notch flexure hinges. *J. Mech. Sci.* **2017**, *8*, 29–49. [CrossRef]

38. Belfiore, N.P.; Broggiato, G.; Verotti, M.; Balucani, M.; Crescenzi, R.; Bagolini, A.; Bellutti, P.; Boscardin, M. Simulation and construction of a MEMS CSFH based microgripper. *Int. J. Mech. Control* **2015**, *16*, 21–30.

39. Cowen, A.; Hardy, B.; Mahadevan, R.; Wilcenski, S. *PolyMUMPs Design Handbook*; MEMSCAP Inc.: Durham, NC, USA, 2011.

40. Cowen, A.; Hames, G.; Monk, D.; Wilcenski, S.; Hardy, B. *SOIMUMPs Design Handbook*; MEMSCAP Inc.: Durham, NC, USA, 2011.

41. Cho, S.W.; Chasiotis, I. Elastic properties and represenative volume element of polycrystalline silicon for MEMS. *Exp. Mech.* **2007**, *47*, 37–49.

42. Hopcroft, M.A.; Nix, W.D.; Kenny, T.W. What is the Young's Modulus of silicon? *J. Microelectromech. Syst.* **2010**, *19*, 229–238. [CrossRef]

43. Boyd, E.J.; Uttamchandani, D. Measurement of the anisotropy of Young's Modulus in single-crystal silicon. *J. Microelectromech. Syst.* **2012**, *21*, 243–249. [CrossRef]

44. Hickey, R.; Kujath, M.; Hubbard, T. Heat transfer analysis and optimization of two-beam microelectromechanical thermal actuators. *J. Vac. Sci. Technol. Vac. Surf. Films* **2002**, *20*, 971–974. [CrossRef]

45. Coutu, R.A.; LaFleur, R.S.; Walton, J.P.K.; Starman, L.A. Thermal management using MEMS bimorph cantilever beams. *Exp. Mech.* **2016**, *56*, 1293–1303. [CrossRef]

46. Zaitsev, B.N. *Blood Cells Study. NT-MDT Spectrum Instruments*; State Research Center of Virology and Biotechnology VECTOR: Koltsovo, Russia, 2015.

47. McConnell, A.D.; Uma, S.; Goodson, K.E. Thermal conductivity of doped polysilicon layers. *J. Microelectromech. Syst.* **2001**, *10*, 360–369. [CrossRef]

48. Kennedy, J.B.; Madugula, M.K.S. *Elastic Analysis of Structures: Classical and Matrix Methods*; Harper & Row: New York, NY, USA, 1990; Chapters 7–9.

49. Miller, D.C.; Boyce, B.L.; Dugger, M.T.; Buchheit, T.E.; Gall, K. Characteristics of a commercially available silicon-on-insulator MEMS material. *Sensors Actuators A Phys.* **2007**, *138*, 130–144. [CrossRef]

50. Kapels, H.; Aigner, R.; Binder, J. Fracture strength and fatigue of polysilicon determined by a novel thermal actuator. *IEEE Trans. Electron Dev.* **2000**, *47*, 1522–1528. [CrossRef]

51. Tsuchiya, T. Tensile testing of silicon thin films. *Fatigue Fracture Eng. Mater. Struct.* **2005**, *28*, 665–674. [CrossRef]

actuators

MDPI

Article

Polymer Microgripper with Autofocusing and Visual Tracking Operations to Grip Particle Moving in Liquid

Ren-Jung Chang * and Yu-Cheng Chien

Department of Mechanical Engineering, National Cheng Kung University, Tainan 70101, Taiwan;
kissme221017@hotmail.com
* Correspondence: rjchang@mail.ncku.edu.tw; Tel.: +886-6-275-7575-2651

Received: 21 May 2018; Accepted: 7 June 2018; Published: 11 June 2018

Abstract: A visual-servo automatic micromanipulating system was developed and tested for gripping the moving microparticle suspended in liquid well. An innovative design of microgripper integrated with flexible arms was utilized to constrain particles in a moving work space. A novel focus function by non-normalized wavelet entropy was proposed and utilized to estimate the depth for the alignment of microgripper tips and moving particle in the same focus plane. An enhanced tracking algorithm, which is based on Polar Coordinate System Similarity, incorporated with template matching, edge detection method, and circular Hough Transform, was implemented. Experimental tests of the manipulation processes from moving gripper to tracking, gripping, transporting, and releasing 30–50 μm Polystyrene particle in 25 °C water were carried out.

Keywords: microgripper; visual servo; moving particle; autofocus; tracking; gripping

1. Introduction

The manipulation of micro objects in liquid environment has become a great challenge in biotechnical engineering [1]. Micromanipulation of artificial particle or biological cell is a crucial technology for acquiring the geometrical, mechanical, electrical and/or chemical properties of micro objects. In addition, micromanipulation technique can be employed for analyzing the interactions between artificial particle and biological object. Micromanipulation in general requires accurate and precise position and/or force operations in holding, picking, placing, pulling, pushing, cutting, and indenting biotechnical micro objects. Regarding biotechnical manipulation, various approaches have been implemented: optic and electric micromanipulation, magnetic micromanipulation, microelectromechanical systems (MEMS) and mechanical micromanipulation [2]. Early research on utilizing micro end effector for biotechnical manipulation was contributed by Carrozza et al. [3], Ok et al. [4], and Arai et al. [5]. For manipulating moving bioparticle or cell, a non-damage holding technique is generally required before proceeding with further micro operation. In the conventional micromanipulation, a pipette is utilized carefully to hold the micro objects in liquid. In a more recent approach, acoustic pressure field is proposed to preposition micro particles in liquid prior to manipulation by utilizing MEMS gripper [6]. To work with biological samples, such as living cells, it must be possible to immerse the end effector tips into liquids. The micromanipulation technology with device dimensions and material properties compatible to biological cell finds important applications in the biomedical manipulation [7,8]. In considering the different requirements of manipulating micro particles in liquid environment, several mechanical microgrippers have been developed for these applications [1,3,6–18]. Regarding the manipulation of micro particles in liquid environment, the research on gripping stationary artificial particle [1,3,6,7,10–12,15] and living cell [3,6,9,13,14,16] has been reported. For the in vivo operations at smaller size scales, untethered microgrippers that can be navigated to grasp living cell or tissue have been reviewed recently [19,20]. When the micro living cell is moving in liquid, the manipulation technique to hold the cell without causing damage needs further development.

Microscopic autofocus operation is an important technique utilized in the detection, tracking, and grasping of micro objects [21–24]. In gripping objects, it is actually one of the most essential techniques to ensure the successful operation. In the literature, the performance of focus functions has been widely investigated and compared [25–31]. Based on the comparison results of focusing different types of images, in general, recommended effective focus functions include energy Laplace [25], normalized variance [30], wavelet-based measure [31], and autocorrelation for fluorescence microscopy [29]. Regarding the technique for tracking particle, it can be traced back to the Particle Tracking Velocimetry (PTV). In the literature, there are several tracking algorithms for PTV ([32], and references therein). In implementing the digital PTV, temporal tracking scheme of matching the same particles in the consecutive time step is the core technique of each PTV algorithm. A tracking algorithm, which is insensitive to experimental parameters and applied to flows subjected to drift and diffusion, is considered in this paper. Although the performance of focusing functions has been widely tested in the aforementioned literature, it is noted that the focusing operation has not been tested by integrating them with visual tracking and grasping operation. Particularly, the automatic alignment of microgripper tips with moving micro particle in the same focus plane is the most essential test to ensure successful gripping operation.

In this paper, innovative design of microgripper, novel focus function, and enhanced tracking algorithm are proposed for gripping micro particle moving in liquid. After the Introduction, Section 2 describes the system design and installation of the micro manipulation system. An innovative design of microgripper integrated with flexible arms is utilized to constrain particles in a moving work space. Section 3 is dedicated to the innovative autofocus function which is utilized to estimate the depth for the alignment of microgripper tips and moving particle in the same focus plane. Section 4 gives an enhanced two-frame scheme for tracking particle under drift and diffusion flow. Section 5 provides experimental test and record of gripping microparticle. Operational processes from moving gripper to tracking, gripping, transporting, and releasing micro particle were carried out in experimental tests. In Section 6, we summarize the present innovative approach of utilizing microgripper to grip moving particle.

2. System Design and Installation

A micromanipulation system installed to realize grasping micro particle with autofocusing and automatic tracking operations is shown in Figure 1. The micromanipulation system consists of Micro Gripper System, Moving Manipulator System, Object Stage, and Autofocusing Stage. The detail design of the four subsystems is described in the following sections.

2.1. Micro Gripper System and Moving Manipulator System

Micro Gripper System was installed on the Moving Manipulator System. The Moving Manipulator System, MMS-77 from Shimadzu (Kyoto, Japan), is an X-Y-Z stage with 0.1 μm resolution, 30 mm stroke for X and Y axes, and 15 mm stroke for Z axis. The maximum moving speed is 3 mm/s. The Micro Gripper System consists of Gripper and Actuator Mechanism as well as Gripper Controller. The Gripper and Actuator Mechanism include micro gripper and actuating mechanism, as shown in Figure 2. The micro gripper is fabricated by utilizing Excimer Laser, Exitech 2000 (Oxford, UK), with mask projection through $10\times$ size reduction by an optical lens. The micro gripper is driven through a connecting rod by piezoelectric actuator, P-820.10. The design with pushing rod is to minimize the geometrical interference between the piezoelectric actuator and liquid well in gripping operation. By employing the present design, the Micro Gripper System is able to grip a micro particle with diameter around 20–90 μm by the input of 0–100 V from the Gripper Controller, PI (Physik Instrumente) E-503.00 (Munich, Germany).

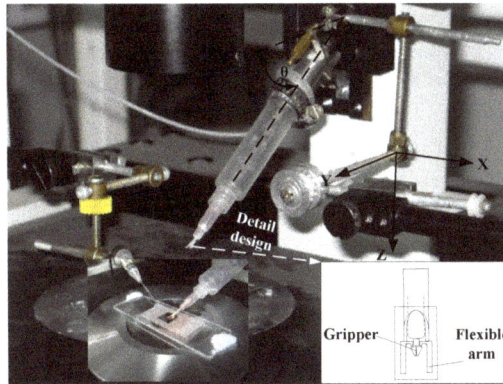

Figure 1. Micromanipulation system (overlapped by a photo of microgripper with bending flexible arms in liquid well).

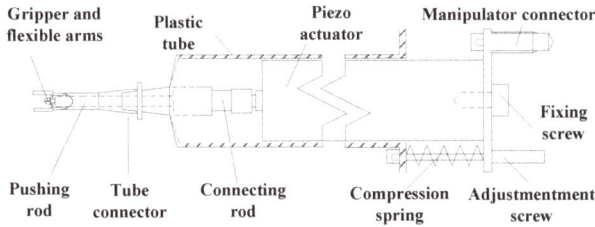

Figure 2. Detail components and assembly of Gripper and Actuator Mechanism.

The innovative micro gripper is made of PU (Polyurethane) film which is 200 μm in thickness. The Young's modulus of PU is 76.4 MPa. The present gripper mechanism is similar to that of the SMA (Shape Memory Alloy) drive PU microgripper employed for gripping micro particle [33]. The mechanism of micro gripper is symmetric with respect to the gripper axis. The kinematic motion of one gripper tip can be modeled through a four-bar orthogonal double slider linkage with one slider along gripper axis as input. Regarding the gripping force, it depends on the geometrical and material property of micro object, gripper–object contact condition, as well as the holding and gripping displacement. When gripping a soft object with Young's modulus less than that of the PU, the gripping force can be analyzed and calibrated in advance through controlling the input displacement. In the present innovative design, as shown in the bottom right corner of Figure 1, the design is to incorporate a pair of flexible arms which are fabricated and integrated on the outer right and left sides of gripper frame. The flexible arms are designed to provide important functions: visual contact probe, constraint of particle movement, and reference for focus plane. When the manipulator moves downward in Z direction, the flexible arms will first approach to touch the inner bottom of liquid container. Then, the continue motion of Moving Manipulator System will cause bending of the flexible arms. The bending flexible arms form a rectangular notch to constrain the fluid flow and trap moving particles. In addition, the reflected edge of bending flexible arms can be monitored through the vertical visual CCD (Charge Coupled Device) system to provide a reference focus plane for positioning the gripper tips and avoiding collision with bottom surface.

2.2. Object Platform

Object Platform is a platform to support particles and fluid container. The platform was installed on another MMS-77 from Shimadzu (Kyoto, Japan). The specifications of MMS-77 to support container are the same as those of MMS-77 employed for installing Micro Gripper System.

2.3. Autofocusing Stage

Autofocusing Stage includes illumination system, CCD visual system, Z-axis servo system, and autofocusing controller. The autofocusing system is implemented through Olympus IX71 (Tokyo, Japan) inverted microscope. With $4\times$ object lens, the view range is $1592 \times 1194 \times 25$ μm^3. The stage is driven by stepping motor with 1.8 degree/step through a gear train of 4.54 ratio. The moving resolution of focusing stage in Z axis by stepping motor is 0.28 μm/degree. The detail components of the image-based Autofocusing Stage are shown in the block diagram of Figure 3. A functional block of the Procedure of autofocusing operation is implemented by PC (Personal computer). The image signal of moving Z position is sensed by CCD through microscope and image capture card. The focusing control through PC can be operated in open-loop or closed-loop mode. The closed-loop control is to realize the autofocusing operation with feedback image signal of moving Z position.

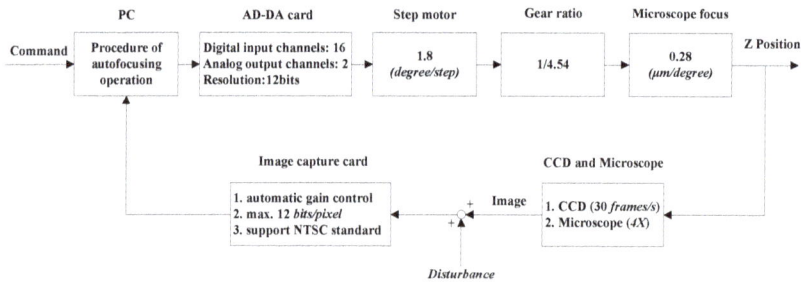

Figure 3. Block diagram of focusing control in open or closed loop through PC.

The synergetic operation of the four subsystems for automatic operation of focusing, tracking and grasping micro particle is implemented by utilizing C/C++ on a PC with Intel® Core™ 2 CPU. The Clock rate of CPU is 2.13 GHz. The human–machine control interface was implemented by employing Microsoft Foundation Classes. The communication between computer and the subsystems is through image capture card, Morphis (MOR/2VD/84) from Matrox (Quebec, Canada), AD/DA cards, PCI-1727U and PCI-1710HG from Advantech (Kaohsiung, Taiwan), and communication interface by RS-232.

3. Autofocusing Operations

A visual servo system with autofocusing operations is an important setup on gripping moving particle. The present image-based autofocusing operation is the first to focus the moving particle and gripper tips, respectively, by evaluating the autofocusing function at different in-focusing positions. Then, the in-focusing positions are utilized to estimate the depth for aligning the gripper tips with moving micro particle in the same focus plane.

Regarding the image-based autofocusing operation, the focus scheme involves searching strategy and evaluating focus measure to locate the focus position. The searching strategy usually includes Global search, Binary search, and Rule-based search [25,34,35]. The selection consideration of searching strategy depends on the searching range, speed, accuracy, and robustness. In the digital searching process, a discrete focus measure which obtains the maximum value in the focusing range is deemed as the best focused image. A discrete focus measure to be formulated is expected to have the properties of the unbiased distribution, clear single apex, high sensitivity, good anti-noise capability, and rapid computation.

In the past, many discrete focus functions have been proposed in the literature. By summing up the different classification of focus functions proposed by Groen et al. [27], Sun et al. [30], and Xie et al. [31], the discrete focus functions in general can be categorized as Derivative-based measure [36–40], Statistics-based measure [27,39,41,42], Histogram-based measure [28], Gray level magnitude-based measure [26,27,29], and Wavelet-based measure [31,43]. In the present focusing and gripping applications, a new focus measure was proposed and the focusing performance was tested and compared with the widely-used measures.

3.1. Wavelet-Entropy Focusing Function

In an image, the diversity of pixel grey level of a focused image is usually greater than that of a defocused one. From the viewpoint of information science, a focusing process can be considered as changes of distribution content of an image. Since an entropy function is a measure of diversity as information content, it may be selected and utilized as a focusing function. Firestone et al. (1991) [28] proposed a histogram-measure Shannon entropy, which is formulated by utilizing the probability of grey level occurred in an image, as a focusing function. The focusing performance of histogram-measure entropy had been tested and compared with other focusing functions. From various comparative studies, the performance of histogram-measure entropy is poor [30,31]. For defining a focus function, it may not be appropriate to formulate entropy through grey level in space domain.

Beyond the information viewpoint on focusing process, a focusing process can be considered as changes of spatial energy content of an image. Application of wavelet energy analysis for defining focus function was proposed by Xie et al. (2007) [31]. The formulation of focus function was to utilize spatial image energy through wavelet coefficients since the orthogonal discrete wavelet transform preserves the contents of image energy. When focusing the image, the energy of high frequency will increase while the energy of low frequency will decrease. In the comparison of four focus measures by Histogram Entropy, Energy Laplace, Normalized Variance, and Wavelet Algorithm, testing results revealed that M_W, which was formulated as a ratio between high frequency energy and low frequency energy, provided the best performance.

The present focusing operation involved two different scales of objects, gripper tips and micro particles, which is different from the aforementioned focusing images under tests. Considering this, a new measure to integrate both information and energy contents in an image may be better utilized for the present applications. First, the Shannon entropy function, which can measure the contents of average information of objects, may be eligible. In the entropy formulation, the discrete Shannon entropy is defined as [28]

$$E_{1D}(X) = -\sum_{i=1}^{n} p_i \ln(p_i) \tag{1}$$

where X is a random variable with n outcomes $\{x_i: i = 1, \ldots, n\}$ and p_i is the probability of outcome x_i. Next, the probability distribution of signal energy relative to the total signal energy can be expressed:

$$p_i = |x_i|^2 / \|X\|^2 \tag{2}$$

where x_i is a series of wavelet coefficients and $\|X\|^2$ is the l^2-norm of $X = \{x_i: i = 1, \ldots, n\}$. The functional in Equation (1) utilizing p_i of Equation (2) in Shannon entropy is non-additive. A one-dimensional additive Shannon entropy may be given as

$$E_{1D}(X) = -\sum_{i=1}^{n} x_i^2 \ln(x_i^2), x_i \neq 0 \tag{3}$$

Here, it is noted that the effect of $\|X\|^2$ is only to shift a function along vertical axis and scale the magnitude of function. Thus, the additive Shannon entropy preserves the function property of Shannon entropy. However, the additive Shannon entropy gives negative value while the Shannon entropy gives positive value.

By following the entropy formulation of Equation (3) and introducing the wavelet energy, the spatial wavelet entropy for a specific wavelet domain may be defined as

$$E^l_{(x,y)\in D} = -\sum_{x=1}^{M}\sum_{y=1}^{N}\left|C^l_{x,y}\right|^2 \log(\left|C^l_{x,y}\right|^2), C^l_{x,y} \neq 0 \tag{4}$$

where D is a wavelet domain on an image and $C^l_{x,y}$ are wavelet coefficients at level l. The image entropy can be expressed as a summation of the entropies contributed by different regions in the wavelet decomposition. By utilizing the mother-wavelet filter of high pass (H) and low pass (L) to an image, the resultant image is divided into four sub-images: LL, HL, LH, snd HH. For level-1 decomposition, the total entropy can be expressed as

$$E^1_{(x,y)\in D} = E^1_{(x,y)\in HH} + E^1_{(x,y)\in HL} + E^1_{(x,y)\in LH} + E^1_{(x,y)\in LL} \tag{5}$$

Since the second moment of wavelet coefficients in the low frequency component will be slightly decreased in the focusing process, this component does not contribute useful focusing information. Consequently, the component of LL in Equation (5) can be ignored and an autofocusing function in the level-1 decomposition, including total high frequency components, is proposed:

$$E_{THF1} = E^1_{(x,y)\in HH} + E^1_{(x,y)\in HL} + E^1_{(x,y)\in LH} \tag{6}$$

For higher level wavelet decomposition, an autofocusing function in the level-2 decomposition can be considered. In considering the issue of real-time computation, only three terms in wavelet decomposition are retained. The contribution of focusing information in the level-2 decomposition is to retain the higher and lower frequency components than those components in the level-1 for obtaining higher contrast between the high frequency and low frequency contents. From Equation (6), an autofocusing function to include significant partial high frequency components in level 2 is proposed:

$$E_{PHF2} = E^2_{(x,y)\in HLHL} + E^2_{(x,y)\in LHLH} + E^2_{(x,y)\in HHHH} \tag{7}$$

The present wavelet-entropy focusing functions are proposed in Equations (6) and (7). The performance of E_{THF1} and E_{PHF2} in the gripping applications needs to be tested.

3.2. Experimental Test and Comparison with Other Focusing Functions

The performance of the proposed autofocusing function on the moving particle was compared with both common focusing functions and wavelet-based focusing functions. In the experimental tests, the particles were made of Polystyrene with density 1.05 g/mL and size 30–50 µm. The microparticles were suspended in water over a long duration. In the experimental tests, the microgripper tips were positioned in the same focus plane as that of microparticles in advance. In utilizing the Autofocusing stage for focusing tests, the block diagram shown in Figure 4 was operated in open loop. The functional block of the Procedure of autofocusing operation in PC was to implement various focusing functions in capturing particle images at different Z positions. The size of particles image was 640 × 480 pixels. The image was captured on every one cyclic rotation of step motor. A total of 29 images were collected. In the sequence of 29 images, Images 1 to 12 were captured in focusing process while Images 13 to 29 were captured in defocusing process. The 29 images in sequence are listed as shown in Figure 4, where the best focusing image is found at image 12.

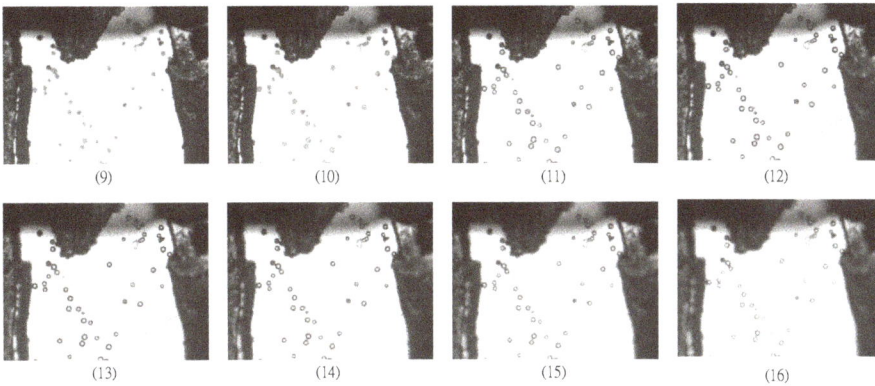

Figure 4. A sequence of eight images from total 29 images are displayed to illustrate the test of autofocusing functions. Images (**1**)–(**12**) are captured in the focusing process while Images (**13**)–(**29**) are captured in the defocusing process. Note that the best focusing image is at Image (**12**).

For the evaluation of noise effect on focusing functions, the relative strength of the added noise is measured by employing *SN* ratio in dB

$$SNR = 10\log_{10} \frac{\frac{1}{M \times N}\sum\limits_{i=1}^{M}\sum\limits_{j=1}^{N}(IM)_{ij}^{2}}{\frac{1}{M \times N}\sum\limits_{i=1}^{M}\sum\limits_{j=1}^{N}((IM)_{ij} - (IMZ)_{ij})^{2}} \tag{8}$$

where $M \times N$ is the size of original image, IM is the grey level of the original image, and IMZ is the grey level of the original image with added noise. In the evaluation of noise effect, a Gaussian white noise with variance 0.0008 was added to each image. The *SNR* of 29 test images was calculated to give 26 ± 0.2 dB.

3.2.1. Comparison with Common Focusing Functions

The focusing performances of the proposed E_{THF1} and E_{PHF2} are compared with the common focusing functions: Normalized variance function, Entropy function, Energy Laplace function, Tenenbaum Gradient function, and Sum Modulus Difference (SMD) function. In the wavelet-based algorithms, the mother wavelet was Haar function, i.e., db2 or Daubechies 2. In comparing performance, the evaluated results of autofocusing functions were normalized, as shown in Figure 5. Figure 5 reveals that all focusing functions give the maximum functional values at the same Z position. The Z position at 12 is a correct in-focus position. However, the autofocusing functions E_{THF1} and E_{PHF2} give sharper curves than those of other autofocusing functions. In addition, the performance of the double decomposition E_{PHF2} is somewhat better than that of single decomposition E_{THF1}.

3.2.2. Comparison of Wavelet-Based Focusing Functions

The autofocusing performance by wavelet-entropy function E_{THF1} and E_{PHF2} was compared with the wavelet-based autofocusing functions by M_w, M1, M2 and M3. Here, the autofocusing functions M1, M2 and M3 correspond to W1, W2, and W3 in [30], respectively. In the comparisons of performance, the evaluated results of autofocusing functions were normalized, as displayed in Figure 6. Figure 6 reveals that all focusing functions give the maximum functional values at the same correct Z position. Among the wavelet-based focusing functions, the present double decomposition E_{PHF2} gives the best performance.

The aforementioned comparison results reveal that the wavelet-entropy function E_{PHF2} gives the best performance in focusing both microparticles and gripper tips. Therefore, the E_{PHF2} is selected and employed for the depth estimation to align microgripper tips and moving particle in the same focus plane.

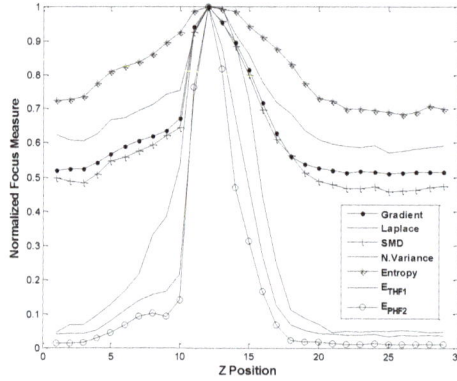

Figure 5. Comparisons of focus measures by wavelet-entropy E_{THF1}, E_{PHF2} and other common focus functions (SNR of test image = 26 ± 0.2 dB).

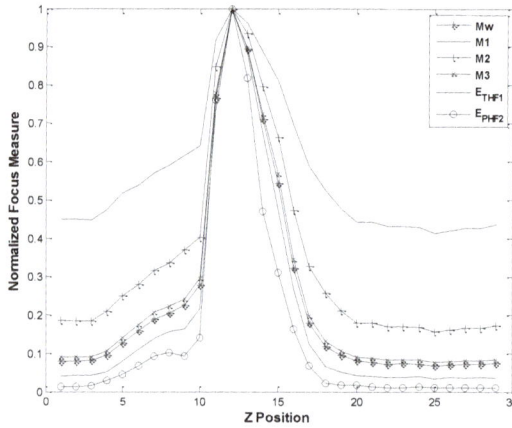

Figure 6. Comparisons of focus measures by wavelet-entropy E_{THF1}, E_{PHF2} and other wavelet-based functions (SNR of test image = 26 ± 0.2 dB).

3.3. Peak Position Identification and Depth Estimation in Visual Servo

3.3.1. Peak Position Identification

In the image-based microscopic autofocus, Global searching strategy was selected for identifying the peak position of focus measure. The image was captured on every one cyclic rotation of step motor. A total of 20 images were collected. The result of E_{PHF2} at different Z positions is shown in Figure 7. Figure 7 shows that there are several local maxima in the focusing function. The highest peak at position 16 refers to the image of the in-focus position of micro particles. The next highest peak at position 3 refers to the image of the in-focus position of gripper tips. To obtain the correct identification of the two local maxima in different Z positions, a template of gripper tips, which was not changed in the process of Global searching, was utilized to distinguish two focusing positions by evaluating the matching score between images. Figure 8 displays the matching score in the process of Global searching. In Figure 8, it is observed that the highest score at position 3 is obviously the in-focus position of gripper tips. The lowest score at position 16 is the in-focus position of microparticles.

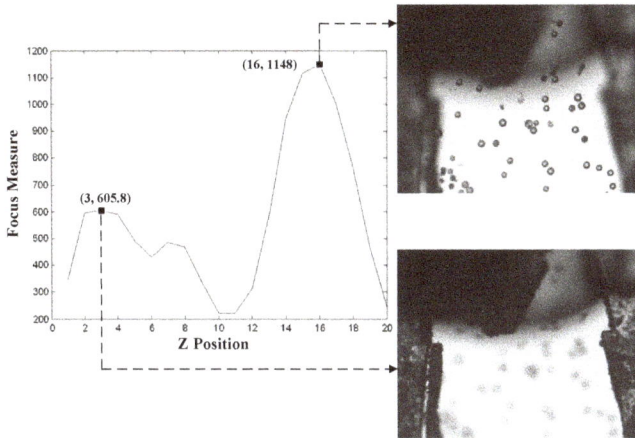

Figure 7. Focus measure of E_{PHF2} in focusing gripper tips and particles.

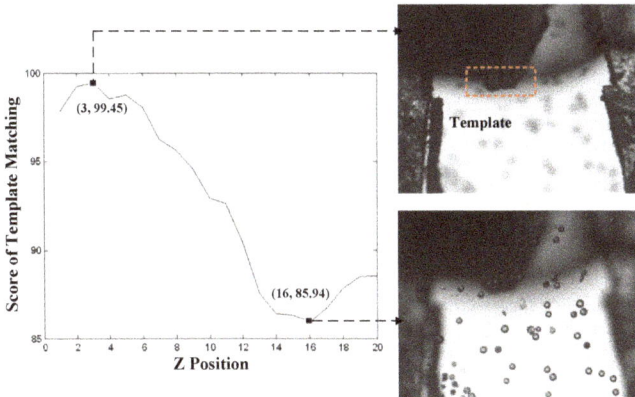

Figure 8. Score of template matching by employing the template of gripper tips.

3.3.2. Depth Estimation

After identifying the in-focus positions of gripper tips and microparticles, the depth between these two positions in Z axis can be evaluated by:

$$Depth = \left| P_{particle} - P_{gripper} \right| \times M_{object\_len} \tag{9}$$

where $P_{particle}$ and $P_{gripper}$ are the Z positions of particles and gripper tip, respectively, and M_{object_len} is the moving distance of object lens by stepping motor in one revolution. Here, $M_{object\_len} = 22.20 \, \mu m/rev.$ is calculated by the multiplication of gain of step motor, gear ratio, and microscope focus from Figure 3. $P_{particle} = 16$ rev. and $P_{gripper} = 3$ rev. are obtained from Figure 8. Consequently, the depth between two focusing plane is calculated to give *Depth* = 288.6 μm. In the servo loop, the depth is calculated in real time for positioning the gripper tips to the same image plane of particles.

The visual servo of autofocusing operation is given by Figure 3. In Figure 3, the functional block of the Procedure of autofocusing operation in PC is to implement the autofocusing operation given in Figure 9. Here, it is noted that the position of gripper tips in Z axis will be moved to the same focusing image plane as that of particles before carrying out the next process of tracking particle.

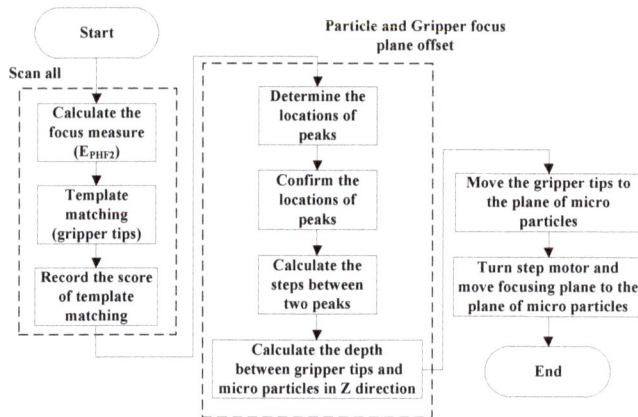

Figure 9. Flowchart of autofocusing operation to align microgripper tips and moving particles in the same focus plane (coarse adjustment).

4. Particle Tracking Process

In utilizing the image-based digital PTV for tracking particle, the first step is to identify and label particles. The next step is to estimate the centroid coordinates of particles. In the last step, the positions of each particle between two consecutive time step images are matched. In considering the requirements of real-time and robust tracking operation, a two-frame scheme by Polar Coordinate System Similarity (PCSS) [44], which is insensitive to experimental parameters and applicable to flows subjected to strong rotation, is first selected and considered for the present applications.

4.1. PCSS Algorithm

The essence of PCSS for finding the matched particles in tracking operation is employing a similarity measure to estimate the changes of both distance and angle of surrounding particles relative to the other particles at two consecutive image frames. The PCSS algorithm is implemented by the following steps:

Step 1. Set an interrogation pattern I centered on the object particle i with radius of R. In the pattern I, the other particles are i_m, where $m = 1, \ldots, M$.

Step 2. Establish a polar coordinate system with the identified centroid of the particle i as the pole. The relative positions of i_m to i is obtained by their polar radii r_{im} and angles θ_{im}.

Step 3. Repeat Steps 1 and 2 for another interrogation pattern J centered on the object particle j with radius of R. In the pattern J, the other particles are j_n, where $n = 1, \ldots, N$, and the relative positions of j_n to j is obtained by their polar radii r_{jn} and angles θ_{jn}.

Step 4. Define a similarity coefficient between patterns I and J as

$$S_{ij} = \sum_{n=1}^{N} \sum_{m=1}^{M} H(\varepsilon_r - |r_{im} - r_{jn}|, \varepsilon_\theta - |\theta_{im} - \theta_{jn}|) \tag{10}$$

$$H(x,y) = \begin{cases} 1, & x > 0, \ y > 0 \\ 0, & otherwise \end{cases} \tag{11}$$

where $H(x,y)$ is a step function. ε_r and ε_θ are thresholds of the polar radius and angle, respectively.

Step 5. Calculate the S_{ij} for all candidate particles in the patterns I and J, and find the matched candidate particle i which gives the maximum value of S_{ij}.

The application of PCSS in tracking particles requires the condition of small velocity change in flow field [32]. However, in considering the present application of tracking micron particle, the particle may fluctuate in Brownian motion which causes low value of S_{ij} in tracking algorithm. Therefore, an enhanced PCSS is proposed to incorporate template matching in the tracking algorithm when the value of S_{ij} is not well above a threshold value ρ.

4.2. Template Matching

The representing particle position in fluctuating movement is obtained through template matching technique. By utilizing zero-mean normalized cross-correlation, the correlation between a template image $w(i,j)$ of size $K \times L$ and an input image $f(x,y)$ of size $M \times N$ can be expressed as

$$\gamma(x,y) = \frac{\sum\limits_{i=0}^{L-1}\sum\limits_{j=0}^{K-1}(w(i,j) - \overline{w})(f_c(x+i,y+j) - \overline{f}_c(x,y))}{\left\{\sum\limits_{i=0}^{L-1}\sum\limits_{j=0}^{K-1}(w(i,j) - \overline{w})^2 \sum\limits_{i=0}^{L-1}\sum\limits_{j=0}^{K-1}(f_c(x+i,y+j) - \overline{f}_c(x,y))^2\right\}^{1/2}} \tag{12}$$

where $\overline{w}$ and $\overline{f}_c$ are the mean of w and f_c, respectively, and the f_c is a subimage of the same size of template. The best match between the template and image is given by the maximum correlation:

$$\gamma(x',y') = \max_{x,y}\{\gamma(x,y)x = 1 \sim M, y = 1 \sim N\} \tag{13}$$

The enhanced PCSS algorithm combines PCSS, edge detection, and template matching in tracking particle movement. In addition, the centroid of particle is estimated by utilizing circular Hough transformation [45]. Regarding the template matching, image pyramids can be employed to enhance the computational efficiency of the correlation-based template detection. The practical operation of the enhanced PCSS algorithm for tracking particle is given by the flowchart shown in Figure 10. In Figure 10, the threshold values are assigned as $\rho = 0.7$, $\varepsilon_r \leq 5$ pixel, and $\varepsilon_\theta \leq 5°$.

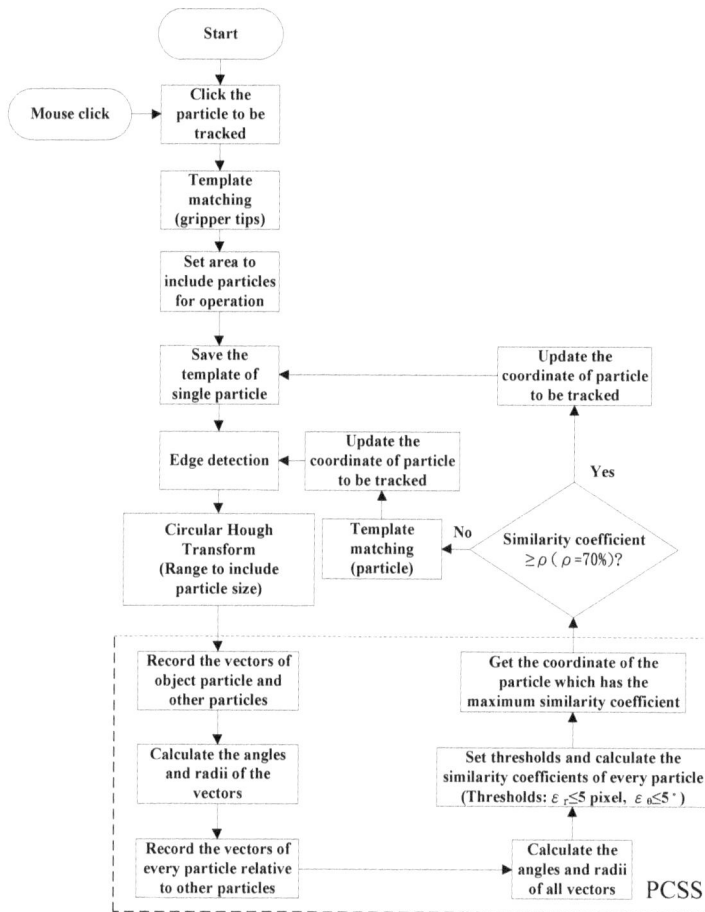

Figure 10. Enhanced PCSS for tracking micro particle on *X-Y* plane.

5. Gripping Microparticle Tests

A visual-servo automatic micromanipulating system was developed for gripping the moving microparticle suspended in liquid well. The performance of immersing the PU gripper tips into water to grip moving Polystyrene particle was tested. In the experimental test, the microgripper was operated with autofocusing and visual tracking functions in gripping operations. The experimental calibration gave $x_{image\_calibration} = 2.494$ μm/pixel and $y_{image\_calibration} = 2.491$ μm/pixel for the calibrated x, y distance per unit pixel, respectively. The size of microparticle was 30–50 μm. The microparticle was suspended in the water well. The water temperature was 25 °C. To emulate the effect of manipulating performance under flow field, the Object Platform to support the water well was moving with speed 0.6 mm/s which was orthogonal to the gripper axis. The moving speed was set through control panel.

5.1. Working Space in Gripping Operation

In the operation of gripping particle, the working space will be established. For the horizontal planner area, it is constrained by the sides of two flexible arms and the line passed through the centroid position of gripper tips' template, as shown in Figure 11. The vertical height is constrained by the 200 μm thickness of microgripper.

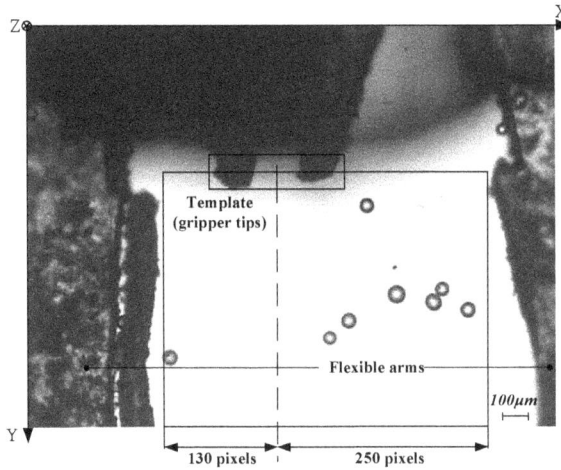

Figure 11. Horizontal planner area to constrain moving particles.

5.2. Fine Adjustment of Tracking in Z axis

Regarding the microparticle suspended in water, the focusing position of individual microparticle can change even though the coarse adjustment in the Z-axis focusing operation is completed. Therefore, fine adjustment of focus position is required for the accurate tracking of the specific particle moving in Z axis. For the fine tracking operation, at first, a planner neighborhood region of the identified particle is established:

$$area\_length = 2\left(R_{\max} + V_x t_{sampling}\right) \qquad (14a)$$

$$area\_width = 2\left(R_{\max} + V_y t_{sampling}\right) \qquad (14b)$$

where $R_{\max}$ is the maximum diameter of particle, $t_{sampling}$ is the sampling time, and V_x and V_y are the estimated particle speed moving in the X and Y direction, respectively. The center of the region is located at the centroid position of the identified particle, which is estimated by the tracking algorithm. Then, a refocusing search is initiated and the focusing measure of E_{PHF2} is recorded at each time step. The fine focusing operation will not stop until the peak value of E_{PHF2} in the next time step is little degraded. The degraded focusing measure can be detected by setting a threshold value δ:

$$\left|\frac{E_{PHF2}(t_i)_{peak} - E_{PHF2}(t_{i+1})_{peak}}{E_{PHF2}(t_i)_{peak}}\right| \leq \delta. \qquad (15)$$

In the experimental tests, $R_{\max} = 20$ pixel, $V_x = V_y = 240$ pixel/s, $t_{sampling} = 0.1$ s, and $\delta = 0.2$.

5.3. Pre-Positioning and Approaching Operations

Before carrying out the operation of gripping identified particle, the microgripper will pre-position itself and then approach to the particle by visual servo. In the first pre-positioning step, the microgripper will track the particle in the horizontal X-Y plane. It will not stop until the centroid of gripper tips' template is located in a region which is close to the identified microparticle. For avoiding collision between gripper tips and particle, a non-colliding safe distance is set to 1.0 $R_{\max}$. The planner region to pre-position the gripper is depicted in Figure 12. The gripper will stop when the following conditions are satisfied:

$$move\_x(t_j) = -(gripper\_x(t_j) - particle\_x(t_j)) \times x_{image\_calibration} \tag{16a}$$

$$\left| mov\_x(t_j) \right| \le 0.3R_{max} \times x_{image\_calibration} \tag{16b}$$

and

$$mov\_y(t_j) = ((particle\_y(t_j) - gripper\_y(t_j))) \times y_{image\_calibration} \tag{16c}$$

$$\left| mov\_y(t_j) - 2R_{max} \times y_{image\_calibration} \right| \le 0.5R_{max} \times y_{image\_calibration} \tag{16d}$$

where $particle\_x(t_j)$, $particle\_y(t_j)$ are the X, Y coordinates of particle position, respectively, $gripper\_x(t_j)$, $gripper\_y(t_j)$ are the X, Y centroid coordinates of gripper tips' template, respectively, and $mov\_x(t_j)$, $mov\_y(t_j)$ are the calculated distances of moving gripper tips to the particle in X, Y directions, respectively. In the second approaching step, the centroid of gripper tips' template will move to approach the particle. By employing visual servo, the input command for the planner movement of microgripper in X and Y directions can be calculated from Equations (16a) and (16c), respectively. If the particle is not accurately tracked by the gripper, the second approaching step can be interrupted by sending operator's command.

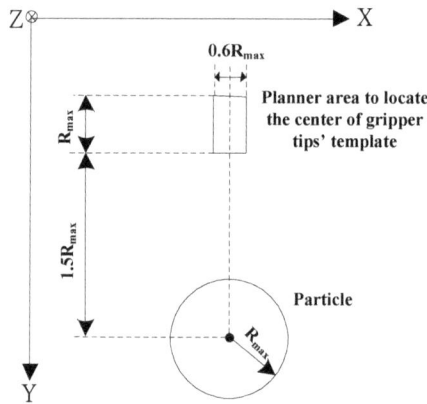

Figure 12. Planner area to locate the gripper by the center of gripper tips' template in pre-positioning operation.

5.4. Gripping and Releasing Operations

For gripping microparticle, the gripper tip is continuously moving to approach the microparticle. It will not move until the position of particle is located near the centroid of gripper tips' template. Then, a gripping command is initiated to close and grip the particle. After gripping particle, the gripper will move upward in $-Z$ direction and then hold the particle to 400 µm away in both X and Y directions. Finally, the gripper tips will move to approach the bottom of well and then the gripper will open for releasing microparticle. To avoid adhesion of the particle to the gripper, the microgripper will move fast along $-Y$ for flushing the particle away.

In the gripping microparticle tests, the operational procedures were: (1) initiating the movement of gripper and capturing gripper tips' template; (2) coarse adjustment; (3) identifying particle by mouse click; (4) fine adjustment; (5) tracking microparticle on X-Y plane; (6) pre-positioning gripper; (7) gripper approaching particle; (8) gripping, transporting, and releasing particle; and (9) homing of gripper. In the initial operation, the microgripper was moved to the position of water surface and the template of gripper tips was captured. During the operational process, the Stop Grasping Command was waited for operator to input if the identified particle was not well tracked by the fine adjustment of focus. The detail operational flowchart is shown in Figure 13. In the performance tests, the images of tracking and gripping identified particle in the 3-D space were captured. The six captured images are displayed in Figure 14. Figure 14(1) shows the initial image in the beginning of global searching for coarse adjustment. Figure 14(2) shows the image of finishing autofocus adjustment. Figure 14(3) displays the image of clicking mouse on the identified particle. Figure 14(4) displays an image in the process of approaching particle before gripping operation. Figure 14(5) displays an image in the process of gripping particle. Figure 14(6) displays an image in the process of releasing particle after the completion of gripping and transporting particle. For the experimental tests, the time expense from the beginning of the global search for coarse adjustment to the actual gripping of the particle was less than 50 s. The successful gripping rate of the present apparatus was 8/10 when the object platform was moving in 0.6 mm/s. The failure was mainly due to the adhesion of micro particles after fine adjustment which caused the inaccurate estimation of particle's coordinates. As a result, the proposed apparatus can perform better when the particle concentration is somewhat low.

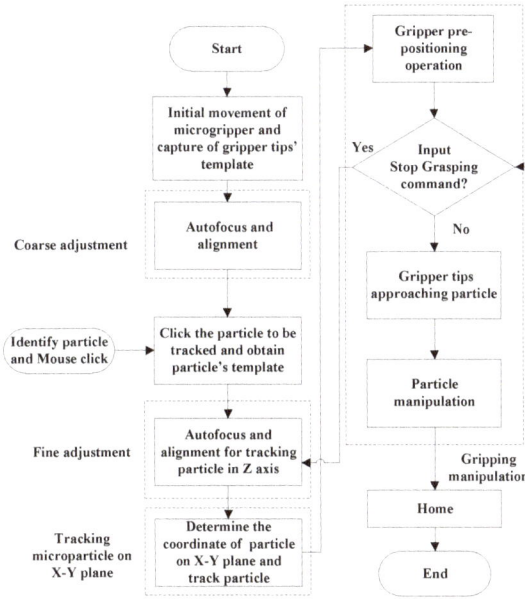

Figure 13. The processes of tracking and gripping particle moving in 3-D space.

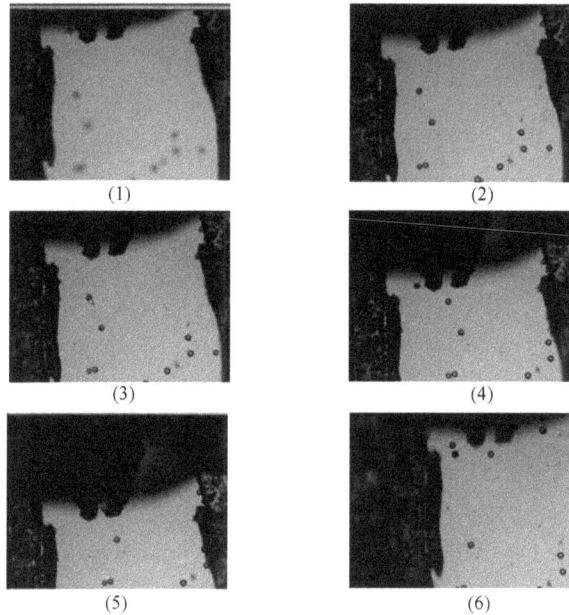

Figure 14. Results of tracking and gripping particle in 3-D space: (**1**) beginning of global searching for coarse adjustment; (**2**) finishing autofocus adjustment; (**3**) identifying particle by clicking mouse; (**4**) approaching particle; (**5**) gripping particle; and (**6**) releasing particle after completion of gripping and transporting particle.

6. Conclusions

An innovative visual-servo automatic micromanipulating system has been developed for gripping the moving microparticle suspended in liquid well. By using an innovative piezoelectric actuated PU micro gripper with flexible arms, a microparticle can be tracked and gripped in a moving work space. A new wavelet-based focus function E_{PHF2}, which revealed the best performance among the surveyed focus functions, was proposed and utilized to estimate the depth for the alignment of microgripper tips and moving particle in the same focus plane. An enhanced PCSS algorithm incorporated with template matching, edge detection method, and circular Hough Transform was implemented for tracking particles under fluctuating and drifting flow. In the performance tests, the images of experimental results in the processes of tracking, gripping, transporting, and releasing 30–50 μm Polystyrene particle in 25 °C water were captured and displayed. For the experimental tests, the time expense from the beginning of the global search for coarse adjustment to the actual gripping of the particle was less than 50 s. The successful gripping rate of the present apparatus was 8/10 when the object platform was moving in 0.6 mm/s. The failure was mainly due to the adhesion of micro particles after fine adjustment which caused the inaccurate estimation of particle's coordinates. As a result, the proposed apparatus can perform better when the particle concentration is somewhat low.

Author Contributions: Methodology, R.J.C.; Writing-Review & Editing, R.J.C.; Software, Y.C.C.; Investigation, R.J.C. and Y.C.C.

Acknowledgments: The authors would like to thank the Ministry of Science and Technology (MOST) for the partial support provided under contract No. MOST 105-2221-E-006-081.

References

1. Jager, E.W.H.; Inganas, O.; Lundstrom, I. Microrobots for micrometer-size objects in aqueous media: Potential tools for single-cell manipulation. *Science* **2000**, *288*, 2335–2338. [CrossRef] [PubMed]
2. Desai, J.P.; Pillarisetti, A.; Brooks, A.D. Engineering approaches to biomanipulation. *Annu. Rev. Biomed. Eng.* **2007**, *9*, 35–53. [CrossRef] [PubMed]
3. Carrozza, M.C.; Dario, P.; Menciassi, A.; Fenu, A. Manipulating biological and mechanical micro-objects using LIGA-microfabricated end-effectors. In Proceedings of the IEEE International Conference on Robotics & Automation, Leuven, Belgium, 16–20 May 1998; Volume 2, pp. 1811–1816.
4. Ok, J.; Chu, M.; Kim, C.J. Pneumatically driven microcage for micro-objects in biological liquid. In Proceedings of the Twelfth IEEE International Conference on Micro Electro Mechanical Systems, Orlando, FL, USA, 17 January 1999; pp. 459–463.
5. Arai, F.; Sugiyama, T.; Luangjarmekorn, P.; Kawaji, A.; Fukuda, T.; Itoigawa, K.; Maeda, A. 3D Viewpoint Selection and Bilateral Control for Bio-Micromanipulation. In Proceedings of the IEEE International Conference on Robotics & Automation, San Francisco, CA, USA, 24–28 April 2000; pp. 947–952.
6. Beyeler, F.; Neild, A.; Oberti, S.; Bell, D.J.; Sun, Y. Monolithically fabricated microgripper with integrated force sensor for manipulating microobjects and biological cells aligned in an ultrasonic field. *J. Microelectromech. Syst.* **2007**, *16*, 7–15. [CrossRef]
7. Feng, L.; Wu, X.; Jiang, Y.; Zhang, D.; Arai, F. Manipulating Microrobots Using Balanced Magnetic and Buoyancy Forces. *Micromachines* **2018**, *9*, 50. [CrossRef]
8. Potrich, C.; Lunelli, L.; Bagolini, A.; Bellutti, P.; Pederzolli, C.; Verotti, M.; Belfiore, N.P. Innovative Silicon Microgrippers for Biomedical Applications: Design, Mechanical Simulation and Evaluation of Protein Fouling. *Actuators* **2018**, *7*, 12. [CrossRef]
9. Chronis, N.; Lee, L.P. Electrothermally activated SU-8 microgripper for single cell manipulation in solution. *J. Microelectromech. Syst.* **2005**, *14*, 857–863. [CrossRef]
10. Hériban, D.; Agnus, J.; Gauthier, M. Micro-manipulation of silicate micro-sized particles for biological applications. In Proceedings of the 5th International Workshop on Microfactories, Becanson, France, 25–27 October 2006; p. 4.
11. Solano, B.; Wood, D. Design and testing of a polymeric microgripper for cell manipulation. *Microelectron. Eng.* **2007**, *84*, 1219–1222. [CrossRef]
12. Han, K.; Lee, S.H.; Moon, W.; Park, J.S.; Moon, C.W. Design and fabrication of the microgripper for manipulation the cell. *Integr. Ferroelectr.* **2007**, *89*, 77–86. [CrossRef]
13. Colinjivadi, K.S.; Lee, J.B.; Draper, R. Viable cell handling with high aspect ratio polymer *chopstick* gripper mounted on a nano precision manipulator. *Microsyst. Technol.* **2008**, *14*, 1627–1633. [CrossRef]
14. Kim, K.; Liu, X.; Zhang, Y.; Sun, Y. Nanonewton force-controlled manipulation of biological cells using a monolithic MEMS microgripper with two-axis force feedback. *J. Micromech. Microeng.* **2008**, *18*, 055013. [CrossRef]
15. Kim, K.; Liu, X.; Zhang, Y.; Cheng, J.; Xiao, Y.W.; Sun, Y. Elastic and viscoelastic characterization of microcapsules for drug delivery using a force-feedback MEMS microgripper. *Biomed. Microdevices* **2009**, *11*, 421–427. [CrossRef] [PubMed]
16. Zhang, R.; Chu, J.; Wang, H.; Chen, Z. A multipurpose electrothermal microgripper for biological micro-manipulation. *Microsyst. Technol.* **2013**, *89*, 89–97. [CrossRef]
17. Di Giamberardino, P.; Bagolini, A.; Bellutti, P.; Rudas, I.J.; Verotti, M.; Botta, F.; Belfiore, N.P. New MEMS tweezers for the viscoelastic characterization of soft materials at the microscale. *Micromachines* **2018**, *9*, 15. [CrossRef]
18. Cauchi, M.; Grech, I.; Mallia, B.; Mollicone, P.; Sammut, N. Analytical, numerical and experimental study of a horizontal electrothermal MEMS microgripper for the deformability characterisation of human red blood cells. *Micromachines* **2018**, *9*, 108. [CrossRef]
19. Malachowski, K.; Jamal, M.; Jin, Q.R.; Polat, B.; Morris, C.J.; Gracias, D.H. Self-folding single cell grippers. *Nano Lett.* **2014**, *14*, 4164–4170. [CrossRef] [PubMed]
20. Ghosh, A.; Yoon, C.; Ongaro, F.; Scheggi, S.; Selaru, F.M.; Misra, S.; Gracias, D.H. Stimuli-responsive soft untethered grippers for drug delivery and robotic surgery. *Front. Mech. Eng.* **2017**, *3*, 7. [CrossRef]

21. Wang, W.H.; Lin, X.Y.; Sun, Y. Contact detection in microrobotic manipulation. *Int. J. Robot. Res.* **2007**, *26*, 821–828. [CrossRef]

22. Inoue, K.; Tanikawa, T.; Arai, T. Micro-manipulation system with a two-fingered micro-hand and its potential application in bioscience. *J. Biotechnol.* **2008**, *133*, 219–224. [CrossRef] [PubMed]

23. Chen, L.; Yang, Z.; Sun, L. Three-dimensional tracking at micro-scale using a single optical microscope. In Proceedings of the International Conference on Intelligent Robotics and Applications, Wuhan, China, 15–17 October 2008; pp. 178–187.

24. Duceux, G.; Tamadazte, B.; Fort-Piat, N.L.; Marchand, E.; Fortier, G.; Dembele, S. Autofocusing-based visual servoing: Application to MEMS micromanipulation. In Proceedings of the 2010 International Symposium on Optomechatronic Technologies (ISOT), Toronto, ON, Canada, 25–27 October 2010; p. 11747002.

25. Subbarao, M.; Tyan, J.K. Selecting the optimal focus measure for autofocusing and depth-from-focus. *IEEE Trans. Pattern Anal. Mach. Intell.* **1998**, *20*, 864–870. [CrossRef]

26. Brenner, J.F.; Dew, B.S.; Horton, J.B.; King, T.; Neurath, P.W.; Selles, W.D. An automated microscope for cytologic research, a preliminary evaluation. *J. Histochem. Cytochem.* **1976**, *24*, 100–111. [CrossRef] [PubMed]

27. Groen, F.C.A.; Young, I.T.; Ligthart, G. A comparison of different focus functions for use in autofocus algorithms. *Cytometry* **1985**, *6*, 81–91. [CrossRef] [PubMed]

28. Firestone, L.; Cook, K.; Culp, K.; Talsania, N.; Preston, K., Jr. Comparison of autofocus methods for automated microscopy. *Cytometry* **1991**, *12*, 195–206. [CrossRef] [PubMed]

29. Santos, A.; De Solorzano, C.O.; Vaquero, J.J.; Peña, J.M.; Malpica, N.; Del Pozo, F. Evaluation of autofocus functions in molecular cytogenetic analysis. *J. Microsc.* **1997**, *188*, 264–272. [CrossRef] [PubMed]

30. Sun, Y.; Duthaler, S.; Nelson, B.J. Autofocusing in computer microscopy: Selecting the optimal focus algorithm. *Microsc. Res. Tech.* **2004**, *65*, 139–149. [CrossRef] [PubMed]

31. Xie, H.; Rong, W.B.; Sun, L.N. Construction and Evaluation of a Wavelet-Based Focus Measure for Microscopy Imaging. *Microsc. Res. Tech.* **2007**, *70*, 987–995. [CrossRef] [PubMed]

32. Shindler, L.; Moroni, M.; Cenedese, A. Spatial–temporal improvements of a two-frame particle-tracking algorithm. *Meas. Sci. Technol.* **2010**, *21*, 1–15. [CrossRef]

33. Chang, R.J.; Lai, Y.H. Design and implementation of micromechatronic systems: SMA drive polymer microgripper. In *Design, Control and Applications of Mechatronic Systems in Engineering*; Sahin, Y., Ed.; InTechOpen: London, UK, 2017.

34. Yao, Y.; Abidi, B.; Doggaz, N.; Abidi, M. Evaluation of sharpness measures and search algorithms for the auto focusing of high-magnification images. *Proc. SPIE* **2006**, *6246*, 62460G. [CrossRef]

35. Kou, C.J.; Chiu, C.H. Improved auto-focus search algorithms for CMOS image-sensing module. *J. Inf. Sci. Eng.* **2011**, *27*, 1377–1393.

36. Mendelsohn, M.L.; Mayall, B.H. Computer-oriented analysis of human chromosomes-III focus. *Comput. Biol. Med.* **1972**, *2*, 137–150. [CrossRef]

37. Krotkov, E. Focusing. *Int. J. Comput. Vis.* **1987**, *1*, 223–237. [CrossRef]

38. Subbarao, M.; Choi, T.S.; Nikzad, A. Focusing techniques. *J. Opt. Eng.* **1993**, *32*, 2824–2836. [CrossRef]

39. Yeo, T.T.E.; Ong, S.H.; Jayasooriah; Sinniah, R. Autofocusing for tissue microscopy. *Image Vis. Comput.* **1993**, *11*, 629–639. [CrossRef]

40. Nayar, S.K.; Nakagawa, Y. Shape from Focus. *IEEE Trans. Pattern Anal. Mach. Intell.* **1994**, *16*, 824–831. [CrossRef]

41. Vollath, D. Automatic focusing by correlative methods. *J. Microsc.* **1987**, *147*, 279–288. [CrossRef]

42. Vollath, D. The influence of the scene parameters and of noise on the behavior of automatic focusing algorithms. *J. Microsc.* **1988**, *151*, 133–146. [CrossRef]

43. Yang, G.; Nelson, B.J. Wavelet-based autofocusing and unsupervised segmentation of microscopic images. In Proceedings of the IEEE/RSJ International Conference on Intelligent Robots and Systems, Las Vegas, NV, USA, 27–31 October 2003; pp. 2143–2148.

44. Ruan, X.D.; Zhao, W.F. A novel particle tracking algorithm using polar coordinate system similarity. *Acta Mech. Sin.* **2005**, *21*, 430–435. [CrossRef]
45. Duda, R.O.; Hart, P.E. Use of the Hough Transformation to detect lines and curves in pictures. *Commun. ACM* **1972**, *15*, 11–15. [CrossRef]

actuators

MDPI

Article

Innovative Silicon Microgrippers for Biomedical Applications: Design, Mechanical Simulation and Evaluation of Protein Fouling

Cristina Potrich [1,2,*], Lorenzo Lunelli [1,2], Alvise Bagolini [3], Pierluigi Bellutti [3], Cecilia Pederzolli [1], Matteo Verotti [4,5] and Nicola Pio Belfiore [6]

[1]   Laboratory of Biomarker Studies and Structure Analysis for Health, Fondazione Bruno Kessler, Trento I-38123, Italy; lunelli@fbk.eu (L.L.); pederzo@fbk.eu (C.P.)
[2]   CNR- Institute of Biophysics, Unit at Trento, Trento I-38123, Italy
[3]   Micro Nano Fabrication and Characterization Facility, Fondazione Bruno Kessler, Trento I-38123, Italy; bagolini@fbk.eu (A.B.); bellutti@fbk.eu (P.B.)
[4]   Department of Industrial Engineering, University of Trento, Trento 38123, Italy; matteo.verotti@unitn.it
[5]   Mechatronics Prototyping Facility, Trentino Sviluppo S.p.A., Rovereto 38068, Italy
[6]   Department of Engineering, Università degli Studi di Roma Tre, via della Vasca Navale 79, Roma 00146, Italy; nicolapio.belfiore@uniroma3.it
*   Correspondence: cpotrich@fbk.eu; Tel.: +39-0461-314605

Received: 8 March 2018; Accepted: 21 March 2018; Published: 24 March 2018

Abstract: The demand of miniaturized, accurate and robust micro-tools for minimally invasive surgery or in general for micro-manipulation, has grown tremendously in recent years. To meet this need, a new-concept comb-driven microgripper was designed and fabricated. Two microgripper prototypes differing for both the number of links and the number of conjugate surface flexure hinges are presented. Their design takes advantage of an innovative concept based on the pseudo-rigid body model, while the study of microgripper mechanical potentialities in different configurations is supported by finite elements' simulations. These microgrippers, realized by the deep reactive-ion etching technology, are intended as micro-tools for tissue or cell manipulation and for minimally invasive surgery; therefore, their biocompatibility in terms of protein fouling was assessed. Serum albumin dissolved in phosphate buffer was selected to mimic the physiological environment and its adsorption on microgrippers was measured. The presented microgrippers demonstrated having great potential as biomedical tools, showing a modest propensity to adsorb proteins, independently from the protein concentration and time of incubation.

Keywords: microgripper; biocompatibility; mini-invasive surgery; protein fouling; cell manipulation

1. Introduction

The demand for micro-manipulation in biomedical applications increased during the last decades. Hundreds of different micro-systems have indeed been reported in the literature [1,2], with different operational strategies [3] and being mainly dedicated to medical or biological applications. Among others, an untethered mobile microgripper controlled by an external electromagnetic coil system has been reported [4], where the manipulation of different microgels was successfully demonstrated with a pick-and-place robotic heterogeneous 3D assembly technique. A few more examples of microgrippers are present in the literature, piezoelectric [5,6] or thermally [7,8] or even magnetically actuated [8–10]. All these systems, however, present quite large dimensions and are not fully tested for working in a physiological environment, as required for biomedical applications.

The development of new microsystems and, in particular, silicon microgrippers for handling cells or tissues at the microscale was therefore pursued in this paper. Since such manipulators have to handle micro-objects, their size should be comparable to the size of the objects being manipulated. A real advantage of micro manipulation can be indeed achieved only if the manipulator is approximately as small as the sample to be manipulated. Starting from a first series of silicon micro-systems obtained by using a classical process based on Reactive-Ion Etching [11,12], a redesign was introduced here in order to fabricate more accurate, new-concept comb-driven microgripper prototypes using Deep Reactive-Ion Etching (DRIE) technology [13]. In this way, a widening in the microsystem design versatility and therefore in the range of potential applications was obtained.

Although the interest for biocompatibility has been widespread in the literature for years, this topic is less explored in relation to microsystems employed as micromanipulators in the biomedical field. In fact, there are few examples about a real use of microsystems for surgical purposes or for in situ biological tissues grasping [2]. Notably, none of these examples consider the possible side-effects such as protein fouling on microscale tools, which could adversely affect the outcome of surgical operations or biopsies. Among examples showing a use of microtools close to the real application, Gultepe et al. [14] report the ex vivo and in vivo tissue excision by microgrippers, which, however, were injected, visualized and recovered with standard equipment. Moreover, some examples of ex vivo micromanipulation in the physiological environment are reported in the literature—for example, the micro-scale compression of hydrogel microcapsules [15] or the micromanipulation of a micro blood vessel and a cyanobacteria cell [16]. The study of Micro Electro-Mechanical Systems (MEMS) microgripper for the deformability characterization of human red blood cells is also reported [17] as well as the study of MEMS tweezers for the viscoelastic characterization of soft materials, like tissues [18]. Besides silicon-based microgrippers, SU-8 microgrippers are reported for single cell manipulation in physiological solution [19]. All of these examples, however, do not consider the complexity of the in vivo environment, such as the presence of a relevant amount of circulating proteins.

In this context, the evaluation of protein adsorption on microdevices devoted to mini-invasive surgery or micromanipulation of cells and tissues could add value for predicting the in vivo performances of such devices. The adsorption of proteins such as albumin on different bulk materials like silicon or aluminum has indeed been reported in the literature (see, for example, [20–22]). Therefore, the study of protein fouling on microgripper surfaces could help in understanding if the delicate mechanism of microgrippers could be affected by the protein deposits that could be formed in physiological conditions. The issue of protein fouling is indeed well known in the field of silicon-based materials, and several authors reported on different methods to avoid protein fouling, as recently revised by Zhang et al. [23]. Among these methods, the functionalization with polyethyleneglycol and its derivatives are widely reported as well as the fabrication of surfaces with nano- or microtopographical features with anti-fouling properties [23].

This paper presents the design and fabrication process of silicon-based micro grippers. The design of microgrippers adopts an innovative concept based on the so-called pseudo–rigid body model (PRBM) [24], while their realization was achieved by the DRIE process. Furthermore, the biocompatibility of microgrippers was assessed in term of protein fouling. To this purpose, the most abundant protein present in blood (i.e., serum albumin) has been selected for the incubation with the developed devices in order to observe the microgripper behavior in a simplified environment similar to the in vivo conditions. For an easy visualization via confocal microscopy, fluorescently-labeled serum albumin was selected and tested at different protein concentrations and different time of incubation.

2. Material and Methods

2.1. Design of Microgrippers

Two different types of microgrippers were designed and tested in this study, both realized by means of bulk microelectromechanical systems (MEMS) technology using silicon on insulator wafer substrates and deep reactive ion etching [13]. The design of the first prototype was developed using a scheme with two degrees of freedom and was inspired by the simplest rotating jaw system, as schematically reported in Figure 1. The PRBM, reported in Figure 1a), corresponds to the compliant structure illustrated in Figure 1b), where two flexure hinges have been introduced in place of the ordinary revolute pairs in A and D. In the figure, the reference system $x - y$ has also been introduced to monitor the motion of the jaw tip point. By using revolute joint-to-flexure substitution, any PRBM (Figure 1a) can be of inspiration to obtain an original compliant structure (Figure 1b). In the case under study, the left- and right-hand side jaws correspond to the rigid links AB and DC (of the ordinary linkage), which rotate around the fixed points A and D, respectively. When a tissue is grasped by the jaws (BC in Figure 1), the overall scheme looks like a four-bar $ABCD$ linkage whose coupler BC represents an elastic medium.

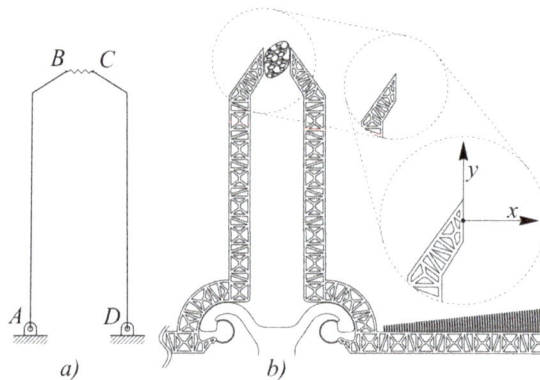

Figure 1. Pseudo-rigid body model (**a**) and original structure (**b**) of the microgripper with two degrees of freedom.

There are several ways of designing a new compliant mechanism, with lumped compliance, starting from a functional representation. The motion of the compliant mechanism can be more or less approximated to the motion of the original ordinary mechanism, depending on the choice of the position of each flexure. In the present investigation, among the several possible ways to transform a functional into a compliant mechanism, two Conjugate Surface Flexure Hinges (CSFH) were the preferred choice to replace the two ordinary revolute joints in A and D. As a consequence, two pseudo–rigid substructures (jaws) were developed in place of the links AB and DC, and a new structure was first developed (see Figure 1b) and then fabricated (see Figure 2). This solution offers the advantage of a large relative rotation (up to $\pm 20°$ using silicon), high accuracy and a reduction of the stress. The advantages of the CSFH rely on the fact that the center of the relative rotations are coincident with the center of the circular beam elastic weights. The microgripper represented in Figure 2 can be operated by finely controlling the torque exerted by the comb-drives, which are attached to the two jaws AB and DC. An accurate estimation of tissue stiffness can then be easily obtained [18].

Considering the viscous behavior of the tissues, however, these microgripper designs, because of their simple structure and of the tissue compliance, can reach some critical configurations. This situation is shown in Figure 3a, where the link BC, representing the visco-elastic tissue, is not rigid anymore and could give rise to the critical configuration depicted in Figure 3b, where the points A, B and C are aligned. From this pose, any arbitrary rotation of link AB can not define a unique rotation because the link DC could be either rotated clockwise or counterclockwise. In this case, a more complicated control strategy should be applied and therefore a more complicated Application Specific Integrated Circuit (ASIC) is required.

As an alternative, a second generation of microgrippers was designed, based on a parallel structure in place of the simple structure with two joint-2 jaws, using a pair of symmetric four-bar linkages, as reported in Figure 4a. This microdevice handles the sample tissue with no intrinsic configuration, all critical configurations being avoided in the neighborhood of the working configuration.

The functional design of the two symmetric four-bar linkages $ABCD$ is illustrated in Figure 4b, where the coupler BC sustains the motion of the coupler point M. A prototype of such microdevice (Figure 5) was fabricated and tested. For clarity, the SEM image presented in Figure 5 also shows lines representing the pair of four-bar linkage $ABCD$ and the coupler point M.

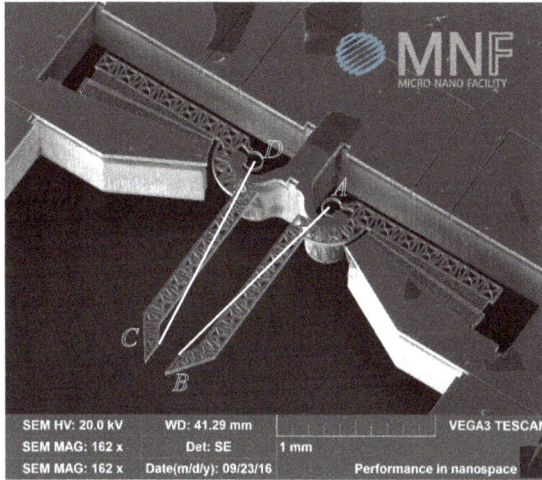

Figure 2. Scanning Electron Microscope (SEM) image of the first microgripper prototype, developed with the simple structure two joint-2 jaws and with two degrees of freedom (white solid lines AB and DC correspond to rigid links, which rotate around the fixed points A and D, respectively).

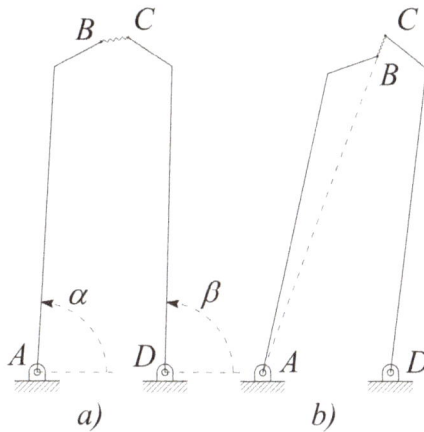

Figure 3. Pseudo-rigid body model of the microgripper with two degrees of freedom in an ordinary (**a**) and a critical (**b**) configuration.

Although this new structure is more complicated than the previous one, having a total of eight CSFHs, both jaws are guided as a rigid link by a loop-closed mechanism and therefore the microsystem has a parallel structure. Furthermore, the rotation center of each jaw is not a fixed point and can be positioned by arranging the mechanism initial configuration. As a consequence, the control strategy for this second microgripper prototype could be simplified with respect to the first prototype described above.

The two microgrippers, illustrated in Figures 2 and 5, indeed have different characteristics and therefore both can be selected depending on whether the application requires a simple layout with limited elastic energy storage in the maximum compressed configuration or a more robust layout with an absence of critical configuration. For these two application cases, the single arm or the four bar linkage type would be the preferred structure, respectively.

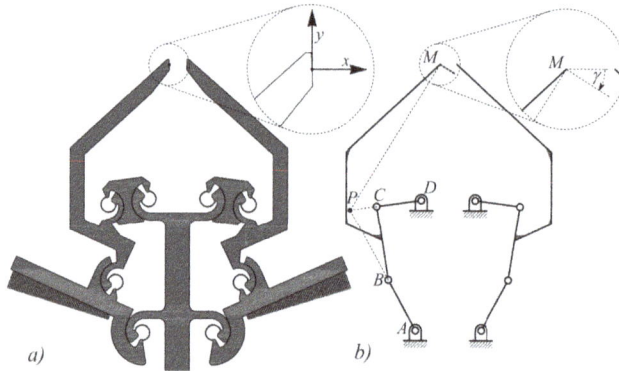

Figure 4. Planar layout (**a**) and functional scheme, PRBM model (**b**), of the second generation microgrippers with a pair of four-bar linkages $ABCD$, each one having four CSFHs and tip M attached to the coupler link BC. Point P represents the instantaneous rotation center for the coupler BC.

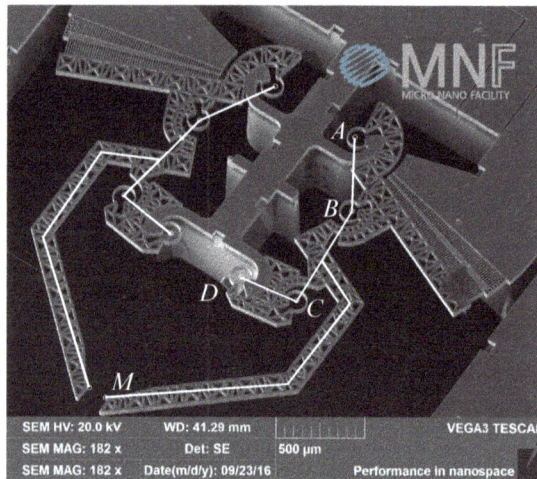

Figure 5. SEM image of a microgripper belonging to the second generation (white solid lines represent two symmetric four-bar linkages, while point M is the coupler point).

2.2. Fabrication of Microgrippers

Fabrication is performed on a 6-inch Semiconductor Equipment and Materials International (SEMI) standard silicon on insulator (SOI) wafers (SOI wafers supplied by SiMat, http://si-mat.com/). The wafer has a 500 μm handle layer that will serve as a support layer for the rigid frames of the device.

On top of it, a 40 μm thick silicon layer is bonded and will provide the device layer, where all moving parts are patterned. A 2 μm oxide layer is in between the two silicon layers, to be used as an etch stop when patterning the silicon and as a support before the complete release of the devices. Each side of the wafer will be patterned, etching the silicon down to the buried oxide layer using the Bosh DRIE process. The fabrication begins with the deposition of a multilayer mask to provide masking for DRIE etching on both sides. First, a 150 nm silicon oxide is deposited, followed by 200 nm Aluminum film sputtering and subsequent 100 nm Titanium sputtering. This layer stack is patterned using standard photolithography. An i-line stepper is used for the front side exposure, while a broadband mask aligner is used for the backside. The stack is then etched in IC standard plasma reactors. Once the masks are patterned, the silicon is etched with DRIE using Bosch patented process in an Alcatel SMS200 etcher (ALCATEL SMS, SPTS Technologies Ltd. Ringland Way, Newport, NP UK). The first etching is performed on the front side, 40 μm deep, and then the second etch is performed on the wafer backside, 500 μm deep. Now, the mechanical structure is fully patterned but still supported by the silicon oxide buried layer: this layer is removed by wet etching in hydrofluoric acid based solution to free the devices. In this step, all residual DRIE mask layers are etched as well. Last, an aluminum layer is formed by physical vapor deposition on the front side, in order to allow for electrical connections devoted to the electrostatic actuation. The details of the fabrication sequence as well as its development are reported in Bagolini et al. [13].

2.3. Protein Fouling Assessment

Protein fouling, i.e., the adsorption and aggregation of proteins on surfaces, was measured by incubating a fluorescent protein on microgrippers. Firstly, microgrippers were gently washed in ultrapure water and cleaned with an argon plasma (1 min, 6.8 W, 2 mbar) to remove organic contaminants. Microgrippers were next incubated with the desired concentration of protein (Albumin from Bovine Serum Tetramethylrhodamine conjugated, BSA-TAMRA, ThermoFisher Scientific, Waltham, MA USA) dissolved in a physiological buffer (DPBS, Sigma, St. Louis, MO USA) spanning different times of incubation. At the end of incubation, microgrippers were gently washed with ultrapure water to remove the excess of protein, dried in air and imaged at the confocal microscope Leica SP5-II (Leica Instruments, Wetzlar, Germany), using an Argon (514 nm) laser line for excitation. All samples were observed utilizing a 20× magnification objective and analyzed with the ImageJ software (version 1.51, Bethesda, MD USA, http://imagej.nih.gov/ij) [25]. Both images and emission spectra (from 520 nm to 650 nm with a step of 5 nm) were acquired.

3. Results and Discussion

The microgrippers presented here were carefully designed and studied for their mechanical potentialities, also simulating their response in different configurations. Two prototypes of microgrippers were realized and investigated. They differ in terms of both number of links and number of CSFHs. The first microgripper is composed of two links and two CSFHs, while the latter is more complex, having a pair of four-bar linkages and eight CSFHs. This last solution allows a more efficient mechanical action, avoiding the critical configurations that can be reached by the first type of microgrippers. Both prototypes were fabricated and tested for their biocompatibility in terms of protein fouling, as a first step to evaluate their possible use as biomedical devices.

3.1. Finite Elements Simulation

Finite Elements Analysis (FEA) simulations were performed to evaluate the static response through electrostatic actuation of the microgrippers by using the commercial Finite Elements Analysis package ANSYS [26]. The rotary comb-drive action was modeled as a series of forces f acting perpendicularly to the free-end section of the movable fingers, each force having magnitude equal to that described in [27]:

$$f = \frac{\epsilon h V^2}{g},$$

(1)

where ϵ is the dielectric constant of the air, V is the applied electric potential difference, g is the radial gap between the fingers, and h is the finger thickness. The common center of the arcs defining the fingers was positioned on the centroid of the flexible element, which is also a limit position for the center of rotation of the flexible element [28]. The design parameters of the comb-drive actuator are listed in Table 1, together with the geometric parameters of the constant-curvature flexure. Multi-step analyses were performed, by increasing the applied voltage with steps of 5 V, from 5 V to the tension corresponding to the maximum angular stroke of the comb-drive, i.e., 4.5 and 5° for the one-link and for the four-bars, respectively. Nonlinearity due to large deflections was considered in the analysis setup, and the anisotropic formulation of elasticity for silicon was considered as described in Hopcroft et al. [29]. The generated meshes were properly refined in the flexible elements and in the fingers of the comb-drives. A detail of one of the generated meshes, corresponding to the four-bars microgripper, is shown in Figure 6.

Table 1. Geometric parameters of comb-drive and flexures.

Geometric Parameter	Values 1-Link	Values 4-Bar
Number of fingers	66	64
Finger angle	6°	7°
Finger initial overlap	1.5°	2°
Angular stroke	4.5°	5°
Finger thickness	40 μm	
Finger width	4 μm	
Finger gap	3 μm	
Flexure width	5 μm	
Flexure thickness	5 μm	
Flexure radius	62.5 μm	
Flexure angle	241°	

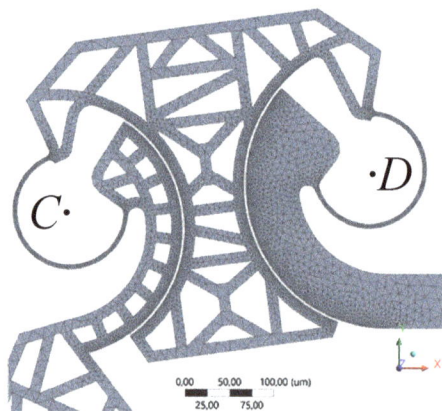

Figure 6. Detail of the refined mesh in two flexures, corresponding to the four-bars microgripper. Letters *C* and *D* refer to the same letters and structures presented in Figure 5.

3.1.1. FEA Simulation of One-Link Microgripper

Microgripper behaviour was simulated through the finite elements analysis, as described in Section 3.1, in order to study their mechanic and kinematic properties for the future use as biomedical microdevices.

For the first prototype, i.e., the one-link microgripper (Figure 2), the generated mesh was composed of 14,401 nodes and 10,563 elements. The maximum angular stroke (4.0°) was reported at the force $f = 28$ μN, corresponding to an applied voltage of 60 V. Figure 7 reports, with respect to the applied electric potential difference, the values of the force exerted by each comb drive and the maximum values of the maximum principal stress (MPS) distribution, determined by means of Equation (1) and FEA, respectively. The maximum value of the MPS is equal to 116 MPa, i.e., below the yield strength of a single-crystal silicon [30]. The path followed by the microgripper tip is depicted in Figure 8. The displacement corresponding to the last simulation step is equal to 103.5 μm along the x-axis and to 17.7 μm along the y-axis.

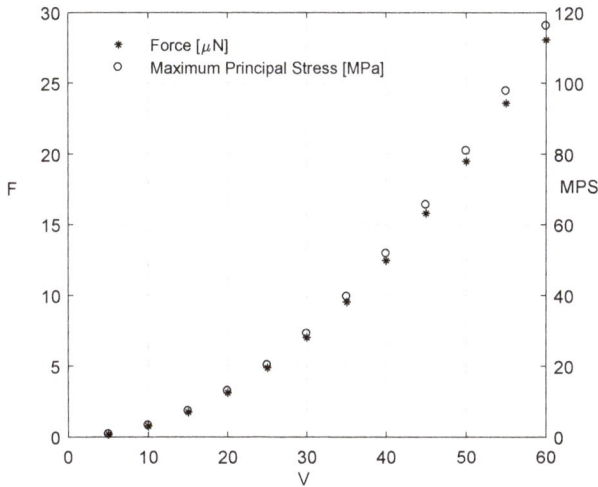

Figure 7. One-link microgripper: force magnitudes exerted by each comb drive and maximum values of the Maximum Principal Stress (MPS) as a function of the applied electric potential difference.

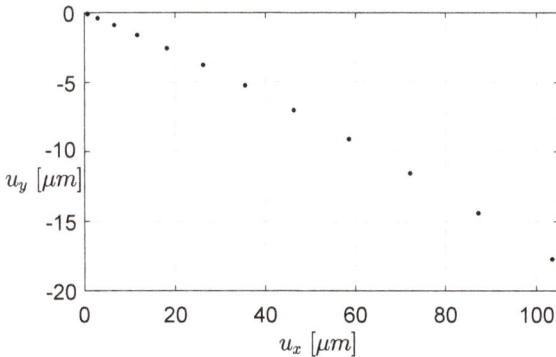

Figure 8. Tip path followed by the one-link microgripper along the x- and y-axes (see Figure 1b).

3.1.2. FEA Simulation of Four-Bar Microgrippers

In the case of the second prototype, i.e., the four-bar microgripper (Figure 5), the generated mesh was composed of 23,345 nodes and 34,990 elements. The maximum angular stroke (4.9°) occurred for f = 317 μN, corresponding to 205 V. Figure 9 reports the values of the exerted force and the MPS distribution. Interestingly, the maximum MPS value is equal to 240 MPa, i.e., below the yield strength of a single-crystal silicon, as already observed for the one-bar microgripper.

The path followed by the microgripper tip is depicted in Figure 10. The displacement corresponding to the last simulation step is equal to 66.2 μm along the x-axis and to 47.4 μm along the y-axis. The simulation of the tip path is particularly important in the case of the four-bar linkage. Considering Figure 4b, the angle γ of inclination of the tangent of the path of the tip point M, in the undeformed configuration, could be easily measured, as evident from the PRBM four-bar linkage. As is already known, angle γ is coincident with the inclination angle of the velocity $\vec{v}_M$ of point M. Therefore, once the center P is identified of instantaneous rotation of the coupler (as made evident in Figure 4b), $\vec{v}_M$, and thus the path tangent in M, must be orthogonal to the line PM. While geometry of the PRBM yields a value of $\gamma = 32°$, Figure 10 shows that the simulated path is tangent to an inclination varying from 32° (at the undeformed initial configuration) to 35° (at the limit position), with a negligible discrepancy both at the beginning and at the extreme position of the path. The two analyses performed for the simulation of microgripper behaviour, i.e., PRBM and FEA, are therefore in good agreement and lead to the conclusion that the method based on the analysis of the PRBM can be successfully employed to design new gripper layouts. This allows designers to conceive new microgrippers that could successfully be employed as biomedical tools for tissue manipulation or for minimally invasive surgery.

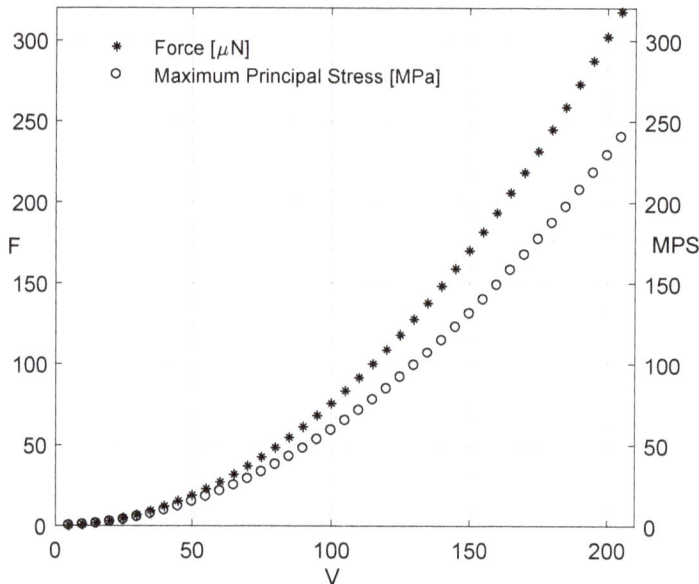

Figure 9. Four-bar microgripper: force magnitudes exerted by each comb drive and maximum values of the MPS as a function of the applied electric potential difference.

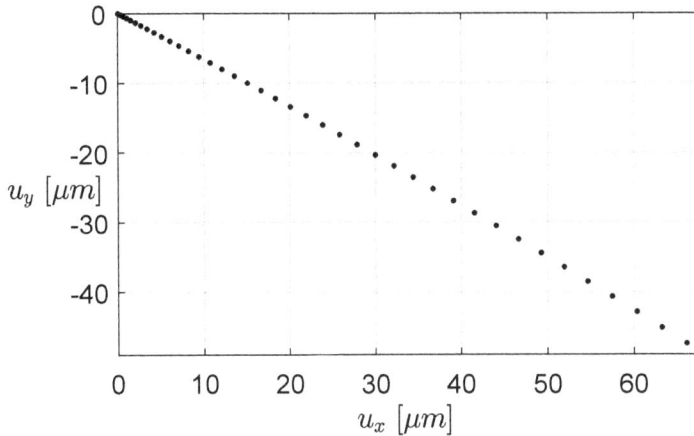

Figure 10. Tip path followed by the four-bar microgripper along the *x*- and *y*-axes (see the Reference system represented in Figure 4a).

3.2. Protein Fouling at the Microgripper Surfaces

Once the good suitability of microgrippers from the mechanic and kinematic point of view is assessed, their compatibility with the use in microsurgery was also investigated. To test the protein propensity to aspecifically aggregate on microgripper surfaces, possibly impeding their correct movements, a test based on protein incubation was selected. The main problem when external objects such as surgical instruments contact body organs or tissues is indeed the adsorption of proteins on such objects [31]. Proteins can easily be adsorbed on external objects, acting as nucleation centers for other materials such as platelets, cells, and bacteria, finally originating a phenomenon known as fouling. The intensity of this phenomenon depends on several parameters, such as the time of contact and the type of surface material. Objects in the microscale range, and in particular their moving parts, could be easily coated by several layers of proteins with the final result of hindering the correct movement of these microdevices. Therefore, the assessment of biocompatibility in terms of protein fouling seems essential for the future use of microgrippers for biomedical applications. Among proteins, serum albumin is the most abundant in human blood, accounting for more than 50% of total proteins [32,33]. The presence of albumin in the extravascular compartment is also significant.

Here, a solution of albumin dissolved in physiological buffer (i.e., DPBS) was selected to mimic the physiological conditions of salts and biomolecules as a simplified but still representative environment. When microgrippers are applied as microtools for picking up tissues for biopsies or as micromanipulators of cultured cells for research purposes, they are in a physiological environment for a non-negligible period of time. During this period, several biological reactions could occur, for example platelet adhesion and activation leading to blood clotting could occur when microgrippers are in contact with blood, or, in the simplest case, proteins could be deposited on the microgripper surface, possibly clogging its delicate mechanisms of motion.

To evaluate the possible effects of protein fouling on microgrippers, the adsorption of different concentrations of a fluorescently-labelled albumin, i.e., BSA-TAMRA, was monitored in time via confocal microscopy. Both sides of microgrippers were observed, since they differ for the external material layer, i.e., silicon or aluminum. The presence of BSA was therefore quantified on both sides in terms of fluorescence intensity per surface area for all the experimental conditions tested. Moreover, both microgripper prototypes were treated and considered in the same way, since the fabrication process was the same as well as microgripper surfaces, i.e., the parts that are in contact with the physiological environment. Considering protein adsorption, the presence of albumin after

30 min incubation was modest on the aluminum side of microgrippers (see Figure 11, panels A and B), while resulting as more evident on the silicon side, as shown in panels C and D of the same figure.

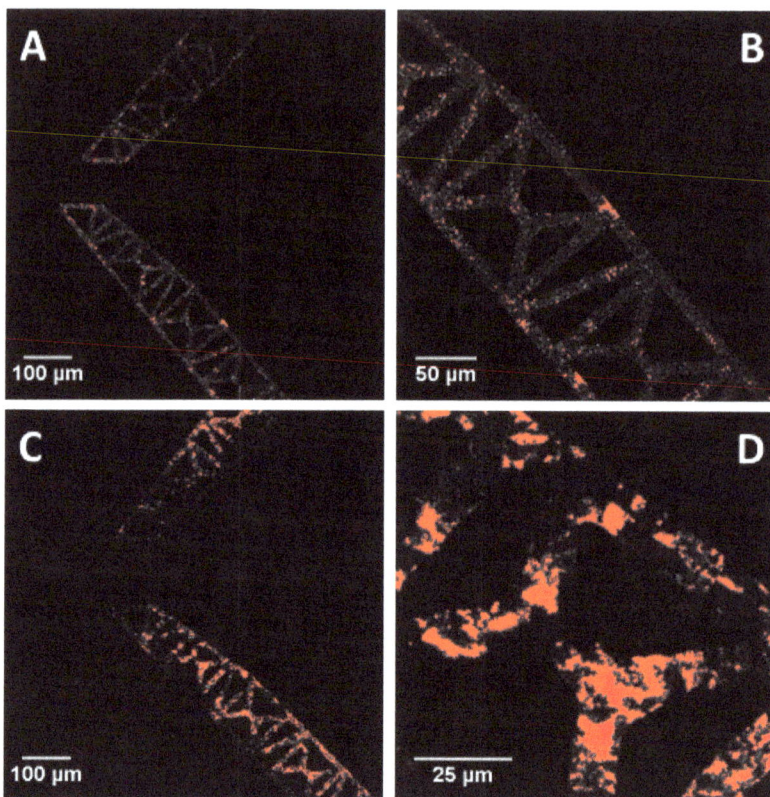

Figure 11. Images of BSA-TAMRA (red regions) adsorbed on the aluminum side of microgrippers (**A** and **B**) or on the silicon side of microgrippers (**C** and **D**), presented with different magnifications. BSA was incubated for 30 min at 0.2 mg/mL, at room temperature. Microgrippers are highlighted in gray.

The same behaviour was observed also after 5 h of BSA incubation, i.e., albumin adsorption was more evident on the silicon side than on the aluminum side (Figure 12, panels B and D). Moreover, the signal acquired on both surfaces was checked for its specificity by measuring the fluorescence spectra on different areas of the images. A clear fluorescence peak corresponding to the maximum emission wavelength of TAMRA (i.e., 580 nm) was always detected from the data corresponding to the red areas (circles in panels A and C of Figure 12), confirming the adsorption of albumin. This conclusion was reinforced by the spectral data taken on the gray areas of the microgrippers (squares in panels A and C of Figure 12), which did not present any specific peak in the fluorescence spectra.

Furthermore, when the emission spectra measured on the two sides of microgrippers were superimposed (Figure 13), exactly the same shape was observed, indicating once more the presence of BSA on both microgrippers' sides.

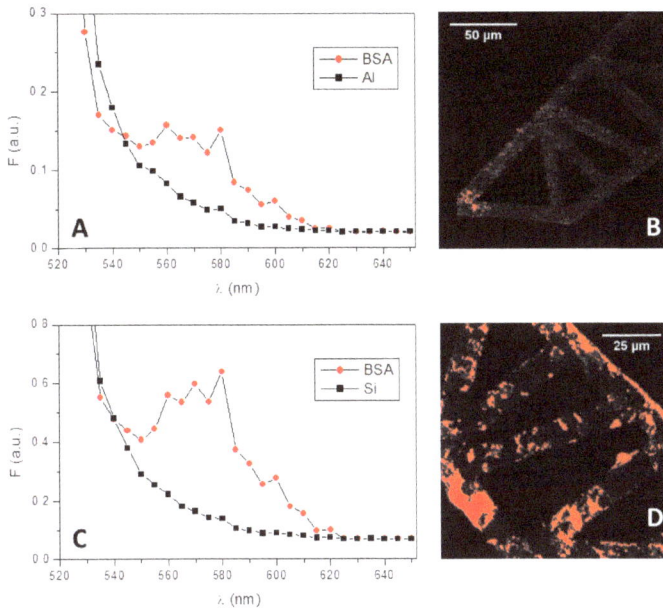

Figure 12. Fluorescence spectra and images of BSA-TAMRA adsorbed on the aluminum side (**A** and **B**) and on the silicon side (**C** and **D**) of microgrippers. In addition, 10 mg/mL of BSA-TAMRA were incubated at room temperature for 5 h separately on both microgrippers' sides. Spectra were acquired where the fluorescence signal was higher (i.e., on the red areas, **B** and **D**) and outside these areas but on microgrippers (gray structures, **b** and **d**), giving, respectively, the red circles' curves and the black squares' curves of **A** and **C**.

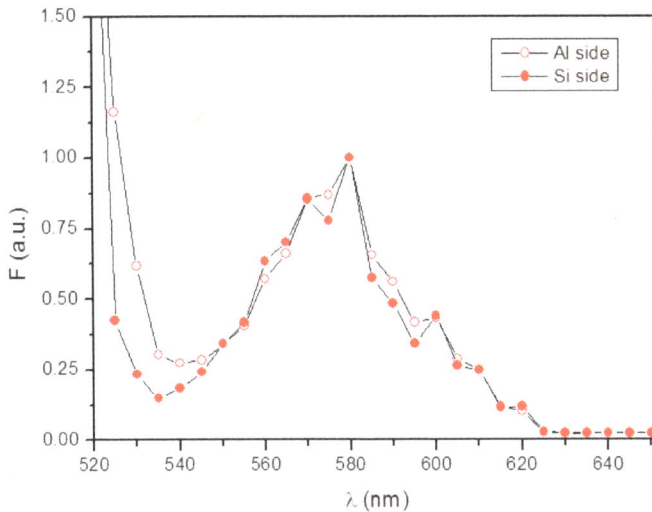

Figure 13. Fluorescence emission spectra measured either on the aluminum side (open circles) or on the silicon side (solid circles) of microgrippers. Spectra were taken on microgrippers treated for 64 h with 0.2 mg/mL BSA-TAMRA at room temperature. Fluorescence intensities are normalized to the maximum wavelength, i.e., 580 nm.

BSA adsorption was also measured as a function of time and concentration (Figure 14). Increasing concentrations of BSA-TAMRA did not significantly alter protein adsorption, which resulted similarly when 0.2 or 10 mg/mL BSA-TAMRA were incubated on both sides of the microgrippers. The adsorption kinetic, however, was faster on the silicon side (panel B, Figure 14) with respect to the aluminum side (panel A, Figure 14). The amount of BSA was indeed higher on silicon even after 30 min incubation, quickly increasing after 5 h incubation. On the contrary, a long-term incubation of 64 h did not result in a broader protein coating, the two microgripper sides being similar in terms of BSA adsorption. The permanence of microgrippers in contact with biological fluids can reasonably be considered in the range from minutes to a few hours: here, a long incubation time (i.e., 64 h) was included in order to stress the conditions leading to protein fouling and fully characterize the biocompatible behavior of microgrippers in view of their future usage in medicine or research.

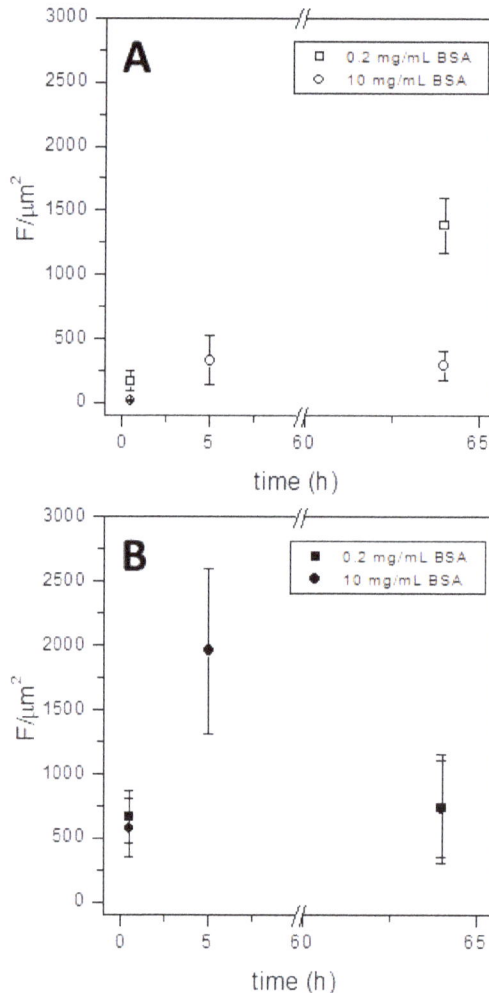

Figure 14. BSA-TAMRA adsorption on the aluminum side (**A**) and the silicon side (**B**) of microgrippers. Adsorption is shown as fluorescence intensity per area unit. Two different concentrations of BSA were incubated, i.e., 0.2 mg/mL (squares) and 10 mg /mL (circles), for 30 min, 5 h or 64 h at room temperature.

Interestingly, a complete protein adsorption has never been observed, neither after the long-term incubation nor at higher BSA concentration. The BSA coating was indeed inhomogeneous also on the silicon side after short incubation time, as witnessed in images of Figures 11 and 12. In other words, microgrippers present a modest propensity to adsorb proteins on both sides. Moreover, this adsorption seems independent of the protein concentration or time of incubation. This property is crucial for microgripper medical applications, such as in mini-invasive surgery, where the clogging mediated by proteins is a non-negligible risk.

4. Conclusions

The complete development of two microgripper prototypes for biomedical applications, fabricated with the MEMS technology and equipped with CSFH, is discussed in detail. In particular, the full process is presented, including the design, the simulation, fabrication, and, finally, the biocompatibility assessment.

The version of the device prototypes presented here has an overall size of about 2 mm × 2 mm, which still seems quite huge in comparison with the objects to be manipulated, such as cells. These dimensions, however, could be easily miniaturized to 1 mm × 1 mm, with the real opportunity of developing a professional inspecting instrument for laboratory measurements, surgical endoscopic operations, or diagnosis of cell or tissue deterioration. Furthermore, by adjusting some process parameters, the overall size can be further reduced to 0.5 mm, including ASIC chips, opening up the possibility to employing these minimally-invasive microgrippers for in vivo applications. All of the described stages gave very positive results, which lead to the conclusion that the two presented microgripper designs have great potential and feasibility for biomedical applications. Experiments to test, for example, the gripping force or displacement at the microgripper tip in the presence of proteins and in the physiological environment are planned to reinforce the conclusions of this study also in working conditions. Experiments like these could define the limit amount of proteins tolerated for microgripper movements or the need for an antifouling treatment before their final use in biomedical applications.

In this context, the good biocompatibility observed in terms of protein fouling plays a crucial role in paving the way to a functional use of microgrippers in mini-surgery or for cell and tissue manipulation for both clinical and research purposes.

Author Contributions: Although the paper is the result of genuine cooperative work, the following specific contributions can be stated. C.Potrich, L.Lunelli and C.Pederzolli performed biocompatibility assay and data interpretation; A.Bagolini and P.Bellutti fabricated microdevices; M.Verrotti and N.P.Belfiore designed microdevices and performed the PRBM and FEA analysis. All authors contributed in manuscript writing and figures drawing.

Conflicts of Interest: The authors declare no conflicts of interest.

Abbreviations

ASIC	Application Specific Integrated Circuit
CSFH	Conjugate Surface Flexure Hinge
DPBS	Dulbecco's Phosphate-Buffered Saline
DRIE	Deep Reactive-Ion Etching
FEA	Finite Elements Analysis
MEMS	Micro Electro-Mechanical Systems
MPS	Maximum Principal Stress
PRBM	Pseudo-Rigid Body Model
SEM	Scanning Electron Microscope

References

1. Verotti, M.; Dochshanov, A.; Belfiore, N.P. A comprehensive survey on microgrippers design: Mechanical structure. *J. Mech. Des.* **2017**, *139*, 060801.
2. Tai, K.; El-Sayed, A.R.; Shahriari, M.; Biglarbegian, M.; Mahmud, S. State of the art robotic grippers and applications. *Robotics* **2016**, *5*, 11.
3. Dochshanov, A.; Verotti, M.; Belfiore, N.P. A comprehensive survey on microgrippers design: Operational strategy. *J. Mech. Des.* **2017**, *139*, 070801.
4. Chung, S.E.; Dong, X.; Sitti, M. Three-dimensional heterogeneous assembly of coded microgels using an untethered mobile microgripper. *Lab on a Chip* **2015**, *15*, 1667–1676.
5. Chen, W.; Zhang, X.; Fatikow, S. A novel microgripper hybrid driven by a piezoelectric stack actuator and piezoelectric cantilever actuators. *Rev. Sci. Instrum.* **2016**, *87*, 115003.
6. El-Sayed, A.M.; Abo-Ismail, A.; El-Melegy, M.T.; Hamzaid, N.A.; Osman, N.A.A. Development of a Micro-Gripper Using Piezoelectric Bimorphs. *Sensors* **2013**, *13*, 5826–5840.
7. Daunton, R.; Gallant, A.; Wood, D.; Kataky, R. A thermally actuated microgripper as an electrochemical sensor with the ability to manipulate single cells. *Chem. Commun.* **2011**, *47*, 6446–6448, doi:10.1039/c1cc11904d.
8. Ongaro, F.; Scheggi, S.; Yoon, C.; Van den Brink, F.; Oh, S.H.; Gracias, D.H.; Misra, S. Autonomous planning and control of soft untethered grippers in unstructured environments. *J. Micro-Bio Rob.* **2017**, *12*, 45–52.
9. Ger, T.R.; Huang, H.T.; Chen, W.Y.; Lai, M.F. Magnetically-controllable zigzag structures as cell microgripper. *Lab on a Chip* **2013**, *13*, 2364–2369, doi:10.1039/c3lc50287b.
10. Al Mashagbeh, M.; Al-Dulaimi, T.; Khamesee, M.B. Design and optimization of a novel magnetically-actuated micromanipulator. *Microsyst. Technol.* **2017**, *23*, 3589–3600.
11. Balucani, M.; Belfiore, N.P.; Crescenzi, R.; Verotti, M. The development of a MEMS/NEMS-based 3 D.O.F. compliant micro robot. *Int. J. Mech. Control* **2011**, *12*, 3–10.
12. Belfiore, N.; Broggiato, G.; Verotti, M.; Balucani, M.; Crescenzi, R.; Bagolini, A.; Bellutti, P.; Boscardin, M. Simulation and Construction of a MEMS CSFH Based Microgripper. *Int. J. Mech. Control* **2015**, *16*, 21–30.
13. Bagolini, A.; Ronchin, S.; Bellutti, P.; Chistè, M.; Verotti, M.; Belfiore, N.P. Fabrication of Novel MEMS Microgrippers by Deep Reactive Ion Etching With Metal Hard Mask. *J. Microelectromech. Syst.* **2017**, *26*, 926–934.
14. Gultepe, E.; Randhawa, J.S.; Kadam, S.; Yamanaka, S.; Selaru, F.M.; Shin, E.J.; Kalloo, A.N.; Gracias, D.H. Biopsy with Thermally-Responsive Untethered Microtools. *Adv. Mater.* **2013**, *25*, 514–519.
15. Kim, K.; Liu, X.; Zhang, Y.; Cheng, J.; Wu, X.Y.; Sun, Y. Elastic and viscoelastic characterization of microcapsules for drug delivery using a force-feedback MEMS microgripper. *Biomed. Microdevices* **2009**, *11*, 421–427.
16. Zhang, R.; Chu, J.; Wang, H.; Chen, Z. A multipurpose electrothermal microgripper for biological micro-manipulation. *Microsyst. Technol.* **2013**, *19*, 89–97.
17. Cauchi, M.; Grech, I.; Mallia, B.; Mollicone, P.; Sammut, N. Analytical, Numerical and Experimental Study of a Horizontal Electrothermal MEMS Microgripper for the Deformability Characterisation of Human Red Blood Cells. *Micromachines* **2018**, *9*, 108.
18. Di Giamberardino, P.; Bagolini, A.; Bellutti, P.; Rudas, I.J.; Verotti, M.; Botta, F.; Belfiore, N.P. New MEMS Tweezers for the Viscoelastic Characterization of Soft Materials at the Microscale. *Micromachines* **2017**, *9*, 15.
19. Chronis, N.; Lee, L.P. Electrothermally activated SU-8 microgripper for single cell manipulation in solution. *J. Microelectromech. Syst.* **2005**, *14*, 857–863.
20. Filali, L.; Brahmi, Y.; Sib, J.D.; Bouhekka, A.; Benlakehal, D.; Bouizem, Y.; Kebab, A.; Chahed, L. The effect of amorphous silicon surface hydrogenation on morphology, wettability and its implication on the adsorption of proteins. *Appl. Surf. Sci.* **2016**, *384*, 107–115.
21. Givens, B.E.; Diklich, N.D.; Fiegel, J.; Grassian, V.H. Adsorption of bovine serum albumin on silicon dioxide nanoparticles: Impact of pH on nanoparticle-protein interactions. *Biointerphases* **2017**, *12*, 02D404.
22. Fukuzaki, S.; Urano, H.; Nagata, K. Adsorption of bovine serum albumin onto metal oxide surfaces. *J. Ferment. Bioeng.* **1996**, *81*, 163–167.
23. Zhang, X.; Brodus, D.; Hollimon, V.; Hu, H. A brief review of recent developments in the designs that prevent bio-fouling on silicon and silicon-based materials. *Chem. Cent. J.* **2017**, *11*, 18.
24. Howell, L.; Magleby, S.; Olsen, B. *Handbook of Compliant Mechanisms*; John Wiley and Sons: Hoboken, NJ, USA, 2013.

25. Schindelin, J.; Arganda-Carreras, I.; Frise, E.; Kaynig, V.; Longair, M.; Pietzsch, T.; Preibisch, S.; Rueden, C.; Saalfeld, S.; Schmid, B.; et al. FIJI: An open-source platformfor biological-image analysis. *Nat. Methods* **2012**, *9*, 676–682.

26. ANSYS Inc. 2017. Available online: http://www.ansys.com (accessed on 27 July 2017).

27. Yeh, J.A.; Chen, C.N.; Lui, Y.S. Large rotation actuated by in-plane rotary comb-drives with serpentine spring suspension. *J. Micromech. Microeng.* **2004**, *15*, 201–206.

28. Verotti, M. Analysis of the center of rotation in primitive flexures: Uniform cantilever beams with constant curvature. *Mech. Mach.Theory* **2016**, *97*, 29–50.

29. Hopcroft, M.A.; Nix, W.D.; Kenny, T.W. What is the Young's Modulus of Silicon? *J. Microelectromech. Syst.* **2010**, *19*, 229–238.

30. Petersen, K.E. Silicon as a mechanical material. *Proc. IEEE* **1982**, *70*, 420–457.

31. Anand, G.; Zhang, F.; Linhardt, R.J.; Belfort, G. Protein-Associated Water and Secondary Structure Effect Removal of Blood Proteins from Metallic Substrates. *Langmuir* **2011**, *27*, 1830–1836.

32. Raoufinia, R.; Mota, A.; Keyhanvar, N.; Safari, F.; Shamekhi, S.; Abdolalizadeh, J. Overview of Albumin and Its Purification Methods. *Adv. Pharm. Bull.* **2016**, *6*, 495–507, doi:10.15171/apb.2016.063.

33. Kumar, D.; Banerjee, D. Methods of albumin estimation in clinical biochemistry: Past, present, and future. *Clin. Chim. Acta* **2017**, *469*, 150–160, doi:10.1016/j.cca.2017.04.007.

actuators

MDPI

Article

Modeling a Pull-In Instability in Micro-Machined Hybrid Contactless Suspension

Kirill V. Poletkin * and Jan G. Korvink

The Institute of Microstructure Technology, Karlsruhe Institute of Technology, Hermann-von-Helmholtz-Platz 1, 76344 Eggenstein-Leopoldshafen, Germany; jan.korvink@kit.edu
* Correspondence: kirill.poletkin@kit.edu; Tel.: +49-721-6082-9305

Received: 26 February 2018; Accepted: 16 March 2018; Published: 20 March 2018

Abstract: A micro-machined hybrid contactless suspension, in which a conductive proof mass is inductively levitated within an electrostatic field, is studied. This hybrid suspension has the unique capability to control the stiffness, in particular along the vertical direction, over a wide range, which is limited by a pull-in instability. A prototype of the suspension was micro-fabricated, and the decrease of the vertical component of the stiffness by a factor of 25% was successfully demonstrated. In order to study the pull-in phenomenon of this suspension, an analytical model was developed. Assuming quasi-static behavior of the levitated proof mass, the static and dynamic pull-in of the suspension was comprehensively studied, also yielding a definition for the pull-in parameters of the hybrid suspension.

Keywords: contactless suspension; magnetic levitation; pull-in; MEMS; stability; dynamics

1. Introduction

Micro-machined Contactless Suspensions (μ-CS), employing the phenomena of electromagnetic levitation, eliminate mechanical attachments between stationary and moving parts in Micro-Electro-Mechanical Systems (MEMS). As a result, they provide one solution of a fundamental issue in the micro-world of MEMS related to the domination of friction over inertial forces [1–3]. Through this concept, a new generation of micro-sensors and actuators based on levitation has been demonstrated.

Depending on the source of the force field, μ-CS can be simply classified as electrostatic, magnetic and hybrid [4]. For instance, electrostatic suspensions (μ-ECS) were successfully used in micro-inertial sensors [5–9]. Magnetic suspensions (μ-MCS) can also be further classified as inductive, diamagnetic and superconducting suspensions and have found applications in micro-bearings [10–13], micro-gyroscopes [14–16], micro-accelerometers [17,18], bistable switches [19], nano-force sensors [20,21], manipulation of droplets [22] and solid micro-particles [23,24]. Hybrid suspensions (μ-HCS) combine different force fields, for instance magneto- and electro-static, variable magnetic and electro-static or magneto-static and variable magnetic fields, which represent the main difference between μ-HCS and both μ-ECS and μ-MCS.

In particular, the capabilities of μ-HCS were demonstrated in applications as micro-motors [25], micro-accelerators [26] and micro-gyroscopes [27,28]. A wide range of different operational modes, such as linear and angular positioning, bistable linear and angular actuation and the adjustment of the stiffness components of μ-HCS, were demonstrated and experimentally studied in the prototype reported in [29]. In particular, the stiffness components were adjusted by changing the equilibrium position of the inductively-levitated, disk-shaped proof mass along the vertical axis. Recently, a novel μ-HCS, in which the electrostatic forces acting on the bottom and top surfaces of the inductively-levitated proof mass maintain the equilibrium position of the proof mass and, simultaneously, decrease the vertical component of the stiffness by means of increasing the strength of the electrostatic field, was proposed in [30] and presented in [31]. Thus, μ-HCS establishes a promising direction for further improvement of a range of different micro-sensors and actuators.

Electrostatic actuation is a key principle available to apply force to a passively-levitated micro-object (proof mass) for the adjustment of its static and dynamic characteristics. However, due to the strongly inherent non-linear dependence of the electrostatic forces that act on such a levitating micro-object on its displacement, the stable levitation of a micro-object is restricted by pull-in phenomena [32]. Moreover, due to the fact that the spring constant created by a magnetic suspension of a μ-HCS also has a nonlinear dependence on displacement, the resulting pull-in phenomenon in μ-HCS cannot be described and characterized by the classic pull-in effect occurring in a spring-mass system with only electrostatic actuation [33].

In this work, pull-in phenomena based on the combination of an inductive suspension and electrostatic actuation are analytically and numerically studied in more detail. The qualitative technique developed in [34] to model micro-machined inductive contactless suspensions, where the eddy current within the levitated micro-object is approximated by a magnetic dipole, is used. We note that this method has been recently further generalized in [35,36], where the eddy current is more accurately approximated by a system of dipoles. Once established, a reduced analytical model of the μ-HCS, which describes the behavior of a levitated micro-object in the vertical direction, is developed.

2. Hybrid Suspension

In this section, the fabrication process, as well as the operating principle of the suspension device, including the necessary service electronics for signal processing and preliminary experimental results, are discussed.

2.1. Fabrication

The suspension consists of three structures fabricated independently at the wafer scale, namely a coil structure and the upper and lower electrode structures. These were aligned and assembled into a sandwich by flip-chip bonding into one device with the dimensions: $9.2 \times 9.2 \times 1.74$ mm, as shown in Figure 1. The coil structure consists of two coaxial 3D wire bonded micro-coils similar to those reported in our previous work [11], namely a stabilization and levitation coil, fabricated on a Pyrex substrate using SU-8 2150; see Figure 1b. For this particular device, a height of the coils is 600 μm, and the number of windings is 20 and 12 for the levitation and stabilization coil, respectively, which allow us to stably levitate an aluminum disk-shaped proof mass with a diameter of 3.2 mm and thickness of 30 μm at a levitation height of 150 μm. The bottom electrode structure was fabricated on an SOI wafer having a device layer of 40 μm, a buried oxide of 2 μm and a handling layer of 600 μm. The resistivity of the silicon layer is in the range of $1 \, \Omega$ cm to $30 \, \Omega$ cm. Furthermore, the device layer has a 500 nm oxide layer for passivation, on top of which electrodes are patterned by UV lithography of evaporated Cr/Au layers (20/150 nm), as shown in the left part of Figure 1c. The SU-8 pillars cover the electrodes in order to insulate the proof mass and electrodes and reduce the contact area between the proof mass and the surface, where the proof mass is initially lying flat. The scaled up image at the left of Figure 1c shows the SU-8 pillar having a diameter of 50 μm and a height of 10 μm. After etching the handle layer up to the buried oxide by DRIE, the bottom electrode structure was aligned and bonded onto the coil structure, as shown in Figure 1d.

The top electrode structure was fabricated on a Pyrex substrate. The electrodes patterned by UV lithography of evaporated Cr/Au layers (20/150 nm) as shown in the right part of Figure 1c have the same design as those on the bottom structure. To create a gap between the top and bottom structures, four SU-8 posts of 130 μm in height were fabricated on the top electrode structure. Then, the top structure was aligned and bonded to the bottom one, as shown in Figure 1a.

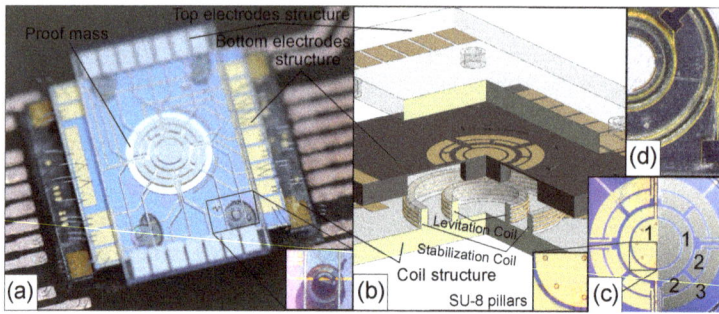

Figure 1. The hybrid suspension: (**a**) the prototype glued to a PCB. The scaled up image at the bottom right corner shows the alignment and the SU-8 post for spacing (the top electrode structure is not connected); (**b**) the exploded view; (**c**) the electrodes patterned at the bottom (right) and top (left); electrode structures: 1, generating negative stiffness; 2, sensing displacement; 3, feedback electrodes; (**d**) a view of the aligning electrode and coil structures from the rear (Pyrex glass) of the device.

In order to avoid using an SOI wafer, and thus also to decrease the amount of parallel capacitance arising in the patterned electrodes due to the conductivity of Si, an alternative fabrication route for the bottom electrode structure was explored based on an intrinsically-doped Si wafer of 500 µm in thickness with a 1-µm oxide layer for passivation, as shown in Figure 2. On one side of the Si wafer, an SU-8 layer of 30 µm to 40 µm in thickness was fabricated by using the epoxy resist SU-8 3025. Then, the electrodes were patterned on this SU-8 layer by UV lithography as shown in Figure 2a. Instead of evaporation, the seed layers Cr/Au (20/150 nm) were sputtered on top of the SU-8 layer. Finally, etching the Si wafer through to the SU-8 layer by DRIE, a cavity for the micro-coils was fabricated as shown in Figure 2b.

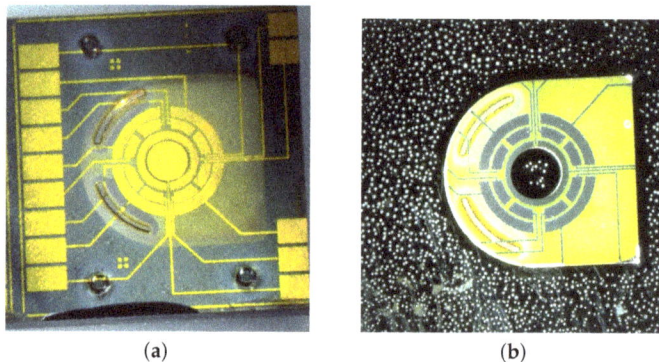

(**a**) (**b**)

Figure 2. The bottom electrode structure fabricated by using a Si wafer with an SU-8 layer of 30 µm in thickness: (**a**) the front side of the structure; (**b**) the rear of the structure.

2.2. Operating Principle

The proof mass is levitated between the electrode structures. A potential U is applied to the top and bottom electrodes (denoted by the number "1") and generates an electrostatic field (see Figure 1c), which causes a decrease of stiffness [30]. The series of electrodes numbered "2" are patterned to realize a differential capacitance for sensing the linear displacement of the proof mass along the vertical axis. Electrodes numbered with "3" generate the electrostatic feedback-force needed to operate in a force-rebalance mode. Thus, the prototype can be considered as a levitated micro-accelerometer operating in the vertical direction and providing an adjustable positional stiffness within closed-loop control.

2.3. Preliminary Experimental Results

In order to provide a proof-of-concept and to demonstrate the successful levitation of the proof mass within the electrostatic field generated by electrodes "1", a preliminary experimental study has been performed (see Figure 3).

Figure 3. The prototype under experimental test: (**a**) the device is fixed on a PCB (front side); (**b**) the interfacial electronics (rear side); (**c**) top, bottom and coil structures are connected to the PCB (scaled image); (**d**) measurements of force against displacement.

To measure the vertical displacement of the proof mass, a circuit for signal processing and conditioning was developed and fabricated. Using the four pairs of electrodes labeled "2" (see Figure 1c), the capacitive sensing for the vertical displacement of the proof mass, based on a capacitance half-bridge and synchronous amplitude demodulation, was implemented. Each electrode of the pairs "2" was excited by an AC voltage having an amplitude of 3 V at a frequency of 100 kHz. After traversing a charge amplifier based on OPA2107AU, the output signal was demodulated by applying a synchronous AM signal. Using switches (ADG441) controlled by a comparator (AD8561), which in turn is synchronized with the excitation voltage and an amplifier (OPA2107AU), a mixer was traversed by the signal. The output from the mixer, passing through an instrumentation amplifier, yielded a differential signal and provided information about the linear displacement.

Coils were fed with a square wave AC current provided by a current amplifier (LCF A093R). The amplitude and frequency of the current in the coils was controlled by a function generator (Arbstudio 1104D) via a computer. A PCB for connecting the top electrodes was fabricated in such way as to leave clear the front of the levitation chip's electrodes, so that a laser beam could reach the proof mass without obstruction, as shown in Figure 3c. This provided us with an additional means to control the linear displacement of the levitated proof mass using a laser distance sensor (LK-G32 with a resolution of 10 nm) and a way to characterize the performance of the capacitive sensing circuit. By applying an electrostatic force generated by the electrodes "3" to the bottom surface of the proof mass, a plot of force against displacement was recorded. From the analysis of the plot, the effective suspension stiffness was estimated.

Assuming that the resulting electrostatic force was applied to the center of the proof mass and accounting for the area of electrode "3" of 4.3×10^{-7} m^2, the electrostatic force generated by the four electrodes was calculated from $F = \varepsilon_0 \varepsilon_r A / 2 \cdot (U/h)^2$, where $\varepsilon_0 = 8.85 \times 10^{-12}$ F m^{-1} is the vacuum permittivity, ε_r is the relative permittivity (for air $\varepsilon_r \approx 1$) and h is the space between an electrode's plane and the equilibrium point of the proof mass.

The results of measurements corresponding to two cases, namely when there is no applied electrical potential to electrodes "1", and when electrodes "1" are energized, are shown in Figure 3d. First, upon energizing electrodes "1", the proof mass was stably levitated, and a decrease of the stiffness from 0.043 to 0.03 N m^{-1} was observed (also see Table 1). Second, a negative stiffness generated by electrodes "1" can be calculated [37] from $NS = -(\varepsilon_0\varepsilon_r A_E U^2)/h^3$, where $A_E = 8.0 \times 10^{-7}$ m^2, to give $NS \approx -0.01$ N m^{-1}. This agrees well with the difference of the two measurements. The results are summarized in Table 1.

Table 1. Parameters of the prototype and experimental results.

Parameters of the Prototype		
Diameter of the proof mass	(mm)	3.2
Thickness of the proof mass	(µm)	30
Levitation height	(µm)	150
Spacing	(µm)	50
Results of Measurements		
Stiffness ($U = 0$)	(N m^{-1})	0.043
Stiffness ($U = 11$ V)	(N m^{-1})	0.03

3. Analytical Model

A schematic diagram for modeling the hybrid contactless suspension is shown in Figure 4a. A typical two-coil stabilization and levitation scheme, arranged to provide stable levitation of a disk-shaped proof mass, is considered. The proof mass is magnetically levitated within the static electric field generated by the top and bottom electrodes. In the general case, it is assumed that the potentials that are applied to the top and bottom electrodes are different and denoted as u_1 and u_2, respectively, as shown in Figure 4a. The equilibrium point coincides with the origin O, which lies on the Z' axis of symmetry. The location of the origin is characterized by the following parameters: h is the spacing between the bottom electrode's plane and the origin, and h_l is the levitation height estimated as the distance between the plane formed by the upper turn of the coils and the origin.

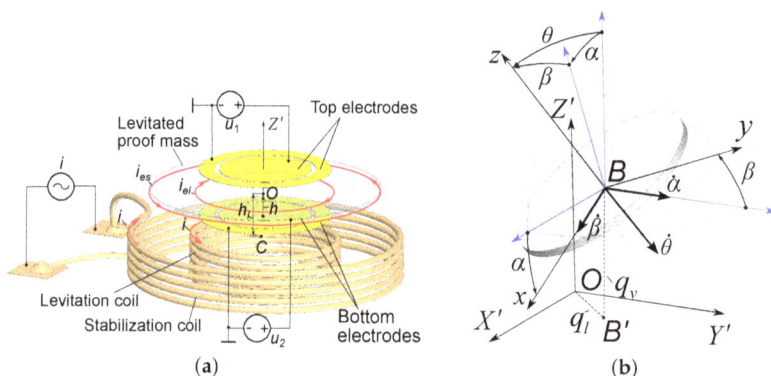

Figure 4. Schematic diagram for modeling the hybrid contactless suspension: (**a**) u_1 and u_2 are the potentials applied to the top and bottom electrodes, respectively; h is the space between an electrode's plane and the equilibrium point of the proof mass; h_l is the levitation height between the plane formed by the upper turn of the coils and the equilibrium point of the proof mass; i_{el} and i_{es} are the eddy currents corresponding to the maximum current density; (**b**) coordinate frames and generalized coordinates to define the position of the disc-shaped proof mass around the origin: q_v, q_l, α and β are the generalized coordinates corresponding to vertical, lateral and angular displacements, respectively.

The behavior of an inductively-levitated disk-shaped proof mass (proof mass) within the static electric field generated by the system of electrodes is strongly non-linear, described by the set of Maxwell equations. However, taking into account the fact that the induced eddy current density within the proof mass is distributed continuously, but not homogeneously, two circuits having maximum values of eddy current density can be identified as the representative circuit for the induced eddy current pattern. Furthermore, assuming quasi-static behavior of the levitated proof mass, a simplification in the mathematical description of the hybrid suspension can be obtained. Applying the qualitative technique proposed in [36], an analytical model of the suspension is formulated. Since the design of the suspension is axially symmetric [36], the mechanical part can be represented by the three generalized coordinates , namely q_v, q_l and θ representing vertical, lateral and angular displacements of the levitated disc, respectively, as introduced in Figure 4b. Considering the capacitors as planar and accounting for $\theta = \alpha + \beta$, the set describing the motion of the hybrid suspended proof mass becomes

$$
\begin{cases}
\dfrac{\partial W_e}{\partial e_1} + \dfrac{\partial \Psi}{\partial \dot{e}_1} = u_1; \quad \dfrac{\partial W_e}{\partial e_2} + \dfrac{\partial \Psi}{\partial \dot{e}_2} = u_2; \\[2mm]
m\ddot{q}_v + \mu_v \dot{q}_v + mg - \dfrac{\partial(W_m - W_e)}{\partial q_v} = F_v; \\[2mm]
m\ddot{q}_l + \mu_l \dot{q}_l - \dfrac{\partial(W_m - W_e)}{\partial q_l} = F_l; \\[2mm]
J\ddot{\theta} + \mu_\theta \dot{\theta} - \dfrac{\partial(W_m - W_e)}{\partial \theta} = T_\theta,
\end{cases}
\tag{1}
$$

where m is the mass, J is the moment of inertia about the axis perpendicular to the disk plane and passing through the center of mass, μ_l, μ_v and μ_θ are the damping coefficients corresponding to the appropriate generalized coordinates, g is the gravity acceleration, F_l, F_v and T_θ are the generalized forces and torque corresponding to the appropriate generalized coordinates, W_m and W_e are energies stored in the magnetic and electric fields, respectively, Ψ is the dissipation function of the system and e_1 and e_2 are the charges on the top and bottom electrodes, respectively. Note that magnetic and electric energies stored in the systems can be described in a way similar to those reported in [30,36].

A necessary, but not sufficient condition for stable levitation of the proof mass, near its equilibrium point, is that the second derivatives of electromagnetic energy stored in the system, defined by the following constants $c_{ij} = -\partial^2(W_m - W_e)/\partial q_i \partial q_j$, where $i = v, l, \theta$ and $j = v, l, \theta$, must correspond to a positive definite quadratic form [38]. Note that the necessary and sufficient conditions for stable levitation in micro-machined inductive suspensions require, in addition, taking into accounting the nonconservative positional force due to the resistivity of the proof mass and the dissipative force acting on the levitated proof mass [36]. Thus, the nonlinear set of Equation (1) forms a generalized analytical model of the hybrid contactless suspension and provides opportunities for modeling its dynamics and stability.

The Accelerometer Equation of Motion

In the framework of the proposed application of the device as an accelerometer, as considered in Section 2.2, the behavior of the proof mass along the vertical direction in the hybrid contactless suspension is of special interest and studied in detail below. The static and dynamic responses of the device along this direction are therefore investigated.

Neglecting the generalized coordinates q_l and q_θ and also assuming that the resistivity of the conducting proof mass and its linear and angular velocities is small, no damping exists and $u_1 = u_2 = U$, then the exact quasi-static nonlinear model, which describes the behavior of the proof mass along the vertical axis, is [30,39]:

$$
m\frac{d^2 q_v}{dt^2} + mg + \frac{I^2}{L}\frac{dM}{dq_v}M - \frac{A}{4}\frac{U^2}{(h - q_v)^2} + \frac{A}{4}\frac{U^2}{(h + q_v)^2} = F_v,
\tag{2}
$$

where I is the amplitude of a harmonic current i in the coils, L is the self-inductance of the proof mass, M is the mutual inductance between the proof mass and coils and U is the applied voltage to the electrodes. Each electrode set has the same area of A_e, $A = \varepsilon_0 A_e$, where ε_0 is the permeability of free space.

In the general case, the mutual inductance M is a complex non-analytical function. This represents the main difficulty for the analytical study of the suspension model (2). However, we can account for some particularities of the micro-machined device, in that the linear sizes of the coils and proof mass are much larger than the levitation height, h_l, and the distribution of the density of the induced eddy current is not homogenous. The induced eddy current is distributed along the levitated proof mass in such a way that two circuits corresponding to maximum values of the eddy current density can be identified, as shown in Figure 4a. The eddy current circuit i_{es} is defined geometrically as a circle having the same diameter as the proof mass. The second circuit i_{el} is also a circle, but with the diameter of the levitation coil [36]. Actually, the second circuit can be considered as the current image of the levitation coil. Due to the mentioned particularities of the device, the force interaction along the vertical direction is reduced to an interaction between eddy current i_{el} and the levitation coil current [40]. Considering both the levitation coil and the eddy current circuit as filamentary circles, the mutual inductance between the levitation coil and eddy current can be described by the Maxwell formula ([41], page 6); thus:

$$k^2 = \frac{4r_l^2}{4r_l^2 + (h_l + y)^2}; \quad M = \mu_0 r_l \left[\left(\frac{2}{k} - k \right) K(k) - \frac{2}{k} E(k) \right], \tag{3}$$

where μ_0 is the magnetic permeability of free space, r_l is the radius of the levitation coil and K and E are complete elliptic integrals of the first and second kinds [42]. Then, accounting for (3), Model (2) becomes:

$$m\frac{d^2 q_v}{dt^2} + mg - \frac{I^2 a^2}{L} \left[\left(\frac{2}{k} - k \right) K(k) - \frac{2}{k} E(k) \right] \frac{2}{k^2} \\ \times \left[\frac{2 - k^2}{2(1 - k^2)} E(k) - K(k) \right] \cdot \frac{\xi^2 (1 + \frac{q_v}{h_l})}{h_l (1 + \xi^2 (1 + \frac{q_v}{h_l})^2)^{3/2}} - \frac{AU^2 q_v}{(h - q_v^2)^2} = F_v, \tag{4}$$

where $a = r_l \mu_0$ and $\xi = h_l/(2r_l)$. Model (4) is analytical, nonlinear and quasi-exact, but due to the elliptic integrals, it can be studied only numerically. For further analysis, Model (4) is presented in dimensionless form as follows:

$$\frac{d^2 \lambda}{d\tau^2} + 1 - \eta \left[\left(\frac{2}{k} - k \right) K(k) - \frac{2}{k} E(k) \right] \frac{2}{k^2} \\ \times \left[\frac{2 - k^2}{2(1 - k^2)} E(k) - K(k) \right] \cdot \frac{\xi^2 (1 + \lambda)\kappa}{(1 + \xi^2 (1 + \lambda)^2)^{3/2}} - \frac{\beta \lambda}{(1 - \lambda^2)^2} = \tilde{F}, \tag{5}$$

where $\tau = \sqrt{g/h} \, t$, $\lambda = q_v/h$, $\eta = I^2 a^2/(mghL)$, $\beta = AU^2/(mgh^2)$, $\kappa = h/h_l$ and $\tilde{F} = F_v/mg$.

Moreover, upon ensuring a condition described further below, Equation (3) can be approximated well by the logarithmic function [34]:

$$M = \mu_0 r_l \left[\ln \frac{4}{\xi(1 + y/h_l)} - 2 \right]. \tag{6}$$

Hence, accounting for the latter equation, the following reduced analytical model of a suspension is proposed:

$$m\frac{d^2 y}{dt^2} + mg - \frac{I^2 a^2}{L} \frac{1}{h_l + y} \left[\ln \frac{4}{\xi(1 + y/h_l)} - 2 \right] - \frac{AU^2 q_v}{(h^2 - q_v^2)^2} = F_v. \tag{7}$$

In dimensionless form, Equation (7) becomes:

$$\frac{d^2\lambda}{d\tau^2} + 1 - \eta \frac{\kappa}{1+\kappa\lambda}\left[\ln\frac{4}{\zeta(1+\kappa\lambda)} - 2\right] - \frac{\beta\lambda}{(1-\lambda^2)^2} = \tilde{F}. \tag{8}$$

As is shown in Appendix A below, the accuracy of approximation of modeling the electromagnetic force is dependent on the parameter ζ. If ζ is less than 0.3, the electromagnetic force is approximated by the logarithmic function (6) with an error less than 6%. When parameter ζ vanishes, the error between the exact Equation (3) and the approximation (6) also vanishes. It is worth noting that, for all known prototypes of μ-HCS published in the literature, parameter ζ is less than 0.25. This fact indicates the applicability of the reduced model for further analytical study of μ-HCS, as has already been successfully demonstrated for instance in [32]. Hence, Model (8) is the main framework for further analysis of the static and dynamic pull-in.

4. Static Pull-In Instability

We now study the load-free behavior of the device upon changing the strength of the electric field, characterized by the dimensionless parameter β (dimensionless squared voltage). For this reason, Equation (8) is written as a set in terms of the phase coordinates [31]:

$$\begin{cases} \dfrac{d\lambda}{d\tau} = \omega; \\ \dfrac{d\omega}{d\tau} = -1 + \eta\dfrac{\kappa}{1+\kappa\lambda}\left[\ln\dfrac{4}{\zeta(1+\kappa\lambda)} - 2\right] + \dfrac{\beta\lambda}{(1-\lambda^2)^2}. \end{cases} \tag{9}$$

From (9), the equilibrium state of the system can be defined as:

$$f(\lambda,\beta) = -1 + \eta\frac{\kappa}{1+\kappa\lambda}[D - \ln(1+\kappa\lambda)] + \frac{\beta\lambda}{(1-\lambda^2)^2}, \tag{10}$$

where $D = \ln\frac{4}{\zeta} - 2$ is the design parameter depending on ζ. At the equilibrium point $\lambda = 0$, the function f must equal zero; this point requires that parameter $\eta = 1/D$. Hence, the static equilibrium state of the system, which relates the vertical coordinate with the strength of the electric field, is:

$$f(\lambda,\beta) = -\frac{\kappa\lambda}{1+\kappa\lambda} - \frac{\ln(1+\kappa\lambda)}{D(1+\kappa\lambda)} + \frac{\beta\lambda}{(1-\lambda^2)^2} \equiv 0. \tag{11}$$

Since the vertical displacement of the proof mass is limited by the positions of the top and bottom electrodes, the variable λ is varied within a range of $-1 \le \lambda \le 1$. Furthermore, taking into account that constant D and κ can be considered within the following ranges of $1 < D < 4.0$ and $0 < \kappa \le 1$, the bifurcation diagram, which relates the distribution of saddles (unstable equilibrium), centers (stable equilibrium) and bifurcations with the dimensionless square voltage β is shown in Figure 5a. Note that the presented prototype of the hybrid suspension in Section 2 has the following values of dimensionless parameters, namely $D = 2.0456$, $\zeta = 0.07$ and $\kappa = 0.3333$. Analysis of the diagram shows that two bifurcation points can be recognized, denoted as A and B. Both bifurcation points correspond to the pull-in instability. This means that, once the strength of the electric field has achieved the value characterized by β_A, the proof mass is pulled in and moves toward the top electrodes. At point B, where the strength of the electric field is characterized by β_B, the proof mass at the position characterized by λ_B is also pulled in, but moves already toward the bottom electrodes.

The bifurcation point A is defined by the following parameters:

$$\lambda_A = 0; \quad \beta_A = \kappa(1+1/D). \tag{12}$$

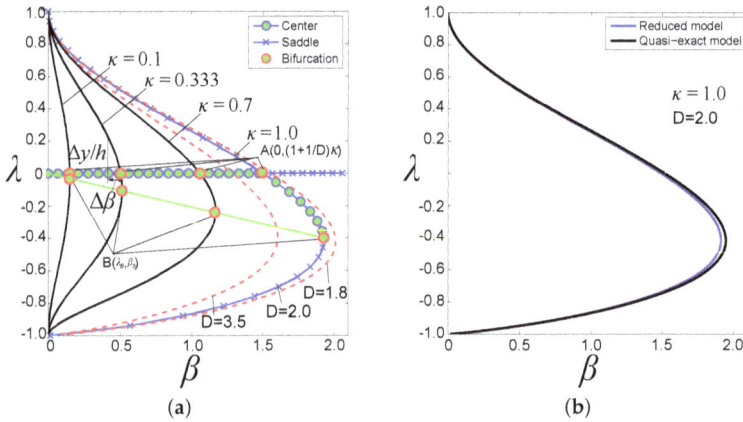

Figure 5. Bifurcation diagram: (**a**) dashed red lines show the evolution of the bifurcation map depending on constant D ($\kappa = 1.0$); solid lines depict the evolution of the bifurcation map depending on spacing $\kappa = h/h_l$ ($D = 2.0$); (**b**) comparison of the quasi-exact and reduced models for $D = 2.0$, $\kappa = 1.0$ and $\zeta = 0.07$ (the relative error is less than 2%).

Parameters β_B and λ_B characterizing bifurcation point B (static pull-in instability) are defined numerically as the solution of the following set of equations:

$$\begin{cases} -3D\kappa^2\lambda^4 - \kappa(4D+1)\lambda^3 - \kappa^2 D\lambda^2 + \kappa\lambda - (2\kappa\lambda^3 + 3\lambda^2 + 2\kappa\lambda + 1)\ln(1 + \kappa\lambda) = 0; \\ \beta = \left(\kappa\lambda + \dfrac{\ln(1+\kappa\lambda)}{D}\right)\dfrac{(1-\lambda^2)}{\lambda(1+\kappa\lambda)}. \end{cases} \quad (13)$$

when κ is small, then the set (13) has an approximate solution:

$$\lambda_B \approx -\frac{\kappa}{4}\frac{D+3/2}{D+1}; \quad \beta_B \approx \kappa(1+1/D)\left(1 + \frac{\kappa^2}{4}\frac{D+3/2}{D+1}\right). \quad (14)$$

In addition, a comparison between reduced Model (8) and the quasi-exact model (5) is performed for the considered design of the hybrid suspension in this work, characterized by the following dimensionless parameters $D = 2$ and $\zeta = 0.07$. The result of this comparison is presented in Figure 5b. Analysis of Figure 5b reveals that the relative error is not in excess of 2%.

The ranges of parameters $\lambda = 0$, $0 \le \beta < \beta_A$ and $-\lambda_B < \lambda < 0$ and $\beta_A \le \beta < \beta_B$ establish a stable state of equilibrium (see Figure 5a). A region near the bifurcation point A is of special interest, because it defines a state of zero stiffness of the suspension. As seen, a decrease of stiffness leads to decreasing a range of linear displacement of the proof mass. Near bifurcation point A, the range of displacement becomes:

$$h\frac{\Delta\beta}{\kappa(1+1/D)} = h\frac{\kappa(1+1/D) - \beta}{\kappa(1+1/D)} \ge \Delta y. \quad (15)$$

Using Equation (15), the minimum possible value of linear stiffness still capable of upholding stable levitation can be estimated. For instance, in the fabricated design of the suspension (see Table 1), upon controlling the linear displacement of the proof mass Δy within a range of ± 1 µm, the relative minimization of the stiffness can be expected to be around 0.007. This means that the initial stiffness generated by the inductive suspension can be reduced by two orders of magnitude. Note that the design of the suspension corresponds to a bifurcation curve with $\kappa = 0.3333$, as shown in Figure 5a.

5. Dynamic Pull-In Instability

An equation for the integral curves of set (9) can be obtained as follows:

$$\frac{d\omega}{d\lambda} = \frac{-\dfrac{\kappa\lambda}{1+\kappa\lambda} - \dfrac{\ln(1+\kappa\lambda)}{D(1+\kappa\lambda)} + \dfrac{\beta\lambda}{(1-\lambda^2)^2}}{\omega}. \tag{16}$$

Integrating (16), the equation of energy is obtained as:

$$\omega^2 + 2\lambda - 2\frac{\ln(1+\kappa\lambda)}{\kappa} + \frac{\ln^2(1+\kappa\lambda)}{D\kappa} - \frac{\beta}{1-\lambda^2} = G, \tag{17}$$

where G is an arbitrary constant of the integration. From the analysis of (17), it is very important to note the following observation, that in order to operate the device properly, it is required to remove the energy of the electric field characterized by parameter β from the system, in order to satisfy the initial conditions. Since G is an arbitrary constant, it can be chosen to be equal to $-\beta$. Then, the final form of the integral equation becomes:

$$\omega^2 + 2\lambda - 2\frac{\ln(1+\kappa\lambda)}{\kappa} + \frac{\ln^2(1+\kappa\lambda)}{D\kappa} - \frac{\beta\lambda^2}{1-\lambda^2} = G'. \tag{18}$$

From (18), the dynamic equilibrium state can be written as:

$$f_d(\lambda, \beta) = 2\lambda - 2\frac{\ln(1+\kappa\lambda)}{\kappa} + \frac{\ln^2(1+\kappa\lambda)}{D\kappa} - \frac{\beta\lambda^2}{1-\lambda^2} \equiv 0. \tag{19}$$

Using (19), the bifurcation diagram can be plotted as shown in Figure 6. Similar to the static bifurcation diagram, it has two pull-in instability points (see Figure 6). One point has the same coordinates as point A shown in (12) corresponding to the static pull-in instability, but B_d has different coordinates compared to the static pull-in point B and can be found by numerically solving the following set:

$$\begin{cases} \dfrac{2\kappa\lambda^2(1-\lambda^2)}{1+\kappa\lambda} - 4\lambda + \left[\dfrac{\kappa\lambda(1-\lambda^2)}{1+\kappa\lambda} + 2D - \ln(1+\kappa\lambda)\right]\dfrac{2}{\kappa D}\ln(1+\kappa\lambda) = 0; \\[3mm] \beta = \left(2\lambda - 2\dfrac{\ln(1+\kappa\lambda)}{\kappa} + \dfrac{\ln^2(1+\kappa\lambda)}{D\kappa}\right)\dfrac{(1-\lambda^2)}{\lambda^2}. \end{cases} \tag{20}$$

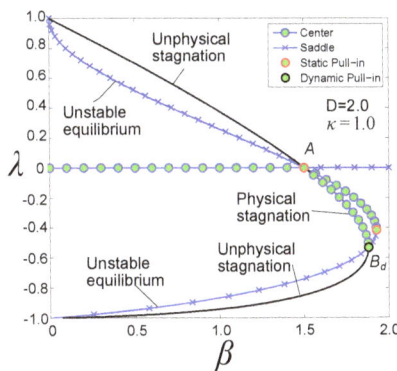

Figure 6. Static and dynamic bifurcation diagrams: solid black lines correspond to unphysical stagnation.

Similar to Section 4, we consider the case when κ is small, then the set (20) has an approximate solution:

$$\lambda_{B_d} \approx -\kappa \frac{D/3 + 1/2}{D+1}; \quad \beta_{B_d} \approx \frac{\kappa}{D}\left(1 - \kappa^2 \frac{(D/3+1/2)^2}{(D+1)^2}\right). \tag{21}$$

Note that, when the spacing κ tends to zero, the static and dynamic pull-in displacements tend to their zero initial position, and pull-in voltages also tend to zero. Once $\kappa = 0$, all static and dynamic pull-in points merge into one zero point.

6. Conclusions

In this article, a micro-machined hybrid suspension based on combining electromagnetic inductive and electrostatic actuation, which provides, in particular, control over and decrease of the vertical component of stiffness, was presented. We discussed the micromachined fabrication process of the device establishing three micro-structures, namely coil, top electrode and bottom electrode structure. In particular, two possible ways of fabrication of the bottom electrode structures based on SOI and Si wafer were considered. Using the developed micro-machined process, a prototype of the suspension was successfully fabricated. The preliminary experimental study of this prototype was performed and successfully demonstrated the proof of concept of the device proposed in [30]. In particular, the disk-shaped aluminum proof mass was levitated between the top and bottom structures generating the electrostatic field. A decrease of the vertical component of the stiffness by 25% was successfully observed.

A generalized analytical model of the suspension was also developed. In order to study the behavior of the hybrid suspension along the vertical direction as a particular case of the developed model, a quasi-exact nonlinear model was formulated. Then, using the design particularities of a micro-machined version of the suspension, a reduced model was proposed. It is worth noting that the applicability of the analytical technique used for modeling a hybrid suspension has been already successfully demonstrated for instance in [32]. Using the reduced nonlinear model, the static and dynamic responses of the suspension were analytically and comprehensively investigated, and the static and dynamic pull-in parameters were identified. In particular, it was shown that within the framework of the developed prototype, the initial stiffness generated by the inductive suspension could be reduced by two orders of magnitude.

Acknowledgments: K.P. acknowledges with thanks the support of the Alexander von Humboldt Foundation of his research project performed at the University of Freiburg. He deeply thanks U. Wallrabe for perfectly hosting and supporting his Humboldt project. K.P. also acknowledges with thanks the support from the German Research Foundation (DFG Grant KO 1883/26-1).

Author Contributions: Kirill V. Poletkin developed and fabricated the device, performed the experiment, developed and studied the models and wrote the manuscript. Jan G. Korvink analyzed intensively the developed model and the obtained experimental results and also contributed in improvement of the writing of the manuscript.

Conflicts of Interest: The authors declare no conflict of interest.

Abbreviations

The following abbreviations are used in this manuscript:

AM	Amplitude modulator
DRIE	Deep reactive-ion etching
NS	Negative stiffness
SIO	Silicon-on-Insulator
μ-CS	Micro-machined Contactless Suspensions
μ-ECS	Micro-machined Electrostatic Suspensions
μ-MCS	Micro-machined Magnetic Suspensions
μ-HCS	Micro-machined Hybrid Suspensions

Appendix A

The accuracy of approximation of modeling the electromagnetic force by means of the reduced model (8) depending on the parameter $\xi = h_l/(2r_l)$, as shown in Figure A1. Analysis of Figure A1 shows that, if the parameter ξ is less than 0.3, the electromagnetic force is approximated by the logarithmic function (6) with an error less than 6%.

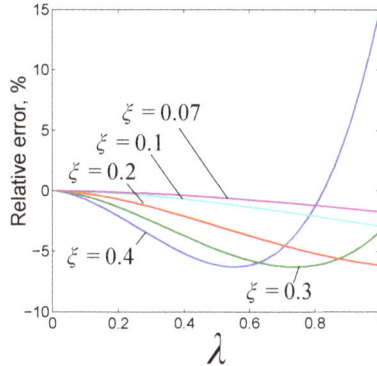

Figure A1. Accuracy of modeling the electromagnetic force by means of the reduced model (8) as compared to the quasi-exact model (5).

References

1. Post, E.R.; Popescu, G.A.; Gershenfeld, N. Inertial measurement with trapped particles: A microdynamical system. *Appl. Phys. Lett.* **2010**, *96*, 143501, doi:10.1063/1.3360808.
2. Poletkin, K.V. Thermal Noise in Ideal Micro-machined Levitated Gyroscope. In Proceedings of the MikroSystemTechnik Congress 2017, Munich, Germany, 23–25 October 2017; VDE Verlag: München, Germany, 2017; pp. 789–792.
3. Poletkin, K.V.; Korvink, J.G.; Badilita, V. Mechanical Thermal Noise in Micro-Machined Levitated Two-Axis Rate Gyroscopes. *IEEE Sens. J.* **2018**, *18*, 1390–1402.
4. Poletkin, K.V.; Badilita, V.; Lu, Z.; Wallrabe, U.; Shearwood, C. *Magnetic Sensors and Devices: Technologies and Applications*; CRC Press: Boca Raton, FL, USA, 2017; pp. 101–131.
5. Toda, R.; Takeda, N.; Murakoshi, T.; Nakamura, S.; Esashi, M. Electrostatically levitated spherical 3-axis accelerometer. Technical Digest. In Proceedings of the Fifteenth IEEE International Conference on Micro Electro Mechanical Systems (Cat. No. 02CH37266), Las Vegas, NV, USA, 24 January 2002; pp. 710–713.
6. Murakoshi, T.; Endo, Y.; Fukatsu, K.; Nakamura, S.; Esashi, M. Electrostatically levitated ring-shaped rotational gyro/accelerometer. *Jpn. J. Appl. Phys.* **2003**, *42*, 2468–2472.
7. Cui, F.; Liu, W.; Chen, W.; Zhang, W.; Wu, X. Design, Fabrication and Levitation Experiments of a Micromachined Electrostatically Suspended Six-Axis Accelerometer. *Sensors* **2011**, *11*, 11206–11234.
8. Han, F.; Liu, Y.; Wang, L.; Ma, G. Micromachined electrostatically suspended gyroscope with a spinning ring-shaped rotor. *J. Micromech. Microeng.* **2012**, *22*, 105032, doi:10.1088/0960-1317/22/10/105032.
9. Han, F.; Sun, B.; Li, L.; Wu, Q. Performance of a sensitive micromachined accelerometer with an electrostatically suspended proof mass. *IEEE Sens. J.* **2015**, *15*, 209–217.
10. Coombs, T.; Samad, I.; Ruiz-Alonso, D.; Tadinada, K. Superconducting micro-bearings. *IEEE Trans. Appl. Supercond.* **2005**, *15*, 2312–2315.
11. Lu, Z.; Poletkin, K.; Hartogh, B.d.; Wallrabe, U.; Badilita, V. 3D micro-machined inductive contactless suspension: Testing and Modeling. *Sens. Actuators A Phys.* **2014**, *220*, 134–143.
12. Poletkin, K.; Moazenzadeh, A.; Mariappan, S.G.; Lu, Z.; Wallrabe, U.; Korvink, J.G.; Badilita, V. Polymer Magnetic Composite Core Boosts Performance of 3D Micromachined Inductive Contactless Suspension. *IEEE Magn. Lett.* **2016**, *7*, 1–3.

13. Poletkin, K.V.; Lu, Z.; Maozenzadeh, A.; Mariappan, S.G.; Korvink, J.G.; Wallrabe, U.; Badilita, V. 3D Micro-machined Inductive Suspensions with the Lowest Energy Consumption. In Proceedings of the MikroSystemTechnik Congress 2017, Munich, Germany, 23–25 October 2017; VDE Verlag: München, Germany, 2017; pp. 500–502.

14. Shearwood, C.; Ho, K.; Williams, C.; Gong, H. Development of a levitated micromotor for application as a gyroscope. *Sens. Actuators A Phys.* **2000**, *83*, 85–92.

15. Su, Y.; Xiao, Z.; Ye, Z.; Takahata, K. Micromachined Graphite Rotor Based on Diamagnetic Levitation. *IEEE Electron Device Lett.* **2015**, *36*, 393–395.

16. Su, Y.; Zhang, K.; Ye, Z.; Xiao, Z.; Takahata, K. Exploration of micro-diamagnetic levitation rotor. *Jpn. J. Appl. Phys.* **2017**, *56*, 126702, doi:10.7567/JJAP.56.126702.

17. Garmire, D.; Choo, H.; Kant, R.; Govindjee, S.; Sequin, C.; Muller, R.; Demmel, J. Diamagnetically levitated MEMS accelerometers. In Proceedings of the International IEEE Solid-State Sensors, Actuators and Microsystems Conference, Lyon, France, 10–14 June 2007; pp. 1203–1206.

18. Ando, B.; Baglio, S.; Marletta, V.; Valastro, A. A Short-Range Inertial Sensor Exploiting Magnetic Levitation and an Inductive Readout Strategy. *IEEE Trans. Instrum. Meas.* **2018**, *PP*, 1–8.

19. Dieppedale, C.; Desloges, B.; Rostaing, H.; Delamare, J.; Cugat, O.; Meunier-Carus, J. Magnetic bistable micro-actuator with integrated permanent magnets. In Proceedings of the IEEE Sensors, Vienna, Austria, 24–27 October 2004; Volume 1, pp. 493–496.

20. Boukallel, M.; Piat, E.; Abadie, J. Passive diamagnetic levitation: theoretical foundations and application to the design of a micro-nano force sensor. In Proceedings of the 2003 IEEE/RSJ International Conference on Intelligent Robots and Systems (IROS 2003) (Cat. No.03CH37453), Las Vegas, NV, USA, 27–31 October 2003; Volume 2, pp. 1062–1067.

21. Abadie, J.; Piat, E.; Oster, S.; Boukallel, M. Modeling and experimentation of a passive low frequency nanoforce sensor based on diamagnetic levitation. *Sens. Actuators A Phys.* **2012**, *173*, 227–237.

22. Lyuksyutov, I.F.; Naugle, D.G.; Rathnayaka, K.D.D. On-chip manipulation of levitated femtodroplets. *Appl. Phys. Lett.* **2004**, *85*, 1817–1819.

23. Chetouani, H.; Jeandey, C.; Haguet, V.; Rostaing, H.; Dieppedale, C.; Reyne, G. Diamagnetic Levitation With Permanent Magnets for Contactless Guiding and Trapping of Microdroplets and Particles in Air and Liquids. *IEEE Trans. Magn.* **2006**, *42*, 3557–3559.

24. Lu, Z.; Chen, P.C.Y.; Lin, W. Force Sensing and Control in Micromanipulation. *IEEE Trans. Syst. Man Cyberne. Part C* **2006**, *36*, 713–724.

25. Liu, W.; Chen, W.Y.; Zhang, W.P.; Huang, X.G.; Zhang, Z.R. Variable-capacitance micromotor with levitated diamagnetic rotor. *Electron. Lett.* **2008**, *44*, 681–683.

26. Sari, I.; Kraft, M. A MEMS Linear Accelerator for Levitated Micro-objects. *Sens. Actuators A Phys.* **2015**, *222*, 15–23.

27. Liu, K.; Zhang, W.; Liu, W.; Chen, W.; Li, K.; Cui, F.; Li, S. An innovative micro-diamagnetic levitation system with coils applied in micro-gyroscope. *Microsyst. Technol.* **2010**, *16*, 431–439.

28. Xu, Y.; Cui, Q.; Kan, R.; Bleuler, H.; Zhou, J. Realization of a Diamagnetically Levitating Rotor Driven by Electrostatic Field. *IEEE/ASME Trans. Mechatron.* **2017**, *22*, 2387–2391.

29. Poletkin, K.; Lu, Z.; Wallrabe, U.; Badilita, V. A New Hybrid Micromachined Contactless Suspension With Linear and Angular Positioning and Adjustable Dynamics. *J. Microelectromech. Syst.* **2015**, *24*, 1248–1250.

30. Poletkin, K.V.; Chernomorsky, A.I.; Shearwood, C. A Proposal for Micromachined Accelerometer, base on a Contactless Suspension with Zero Spring Constant. *IEEE Sens. J.* **2012**, *12*, 2407–2413.

31. Poletkin, K. A novel hybrid contactless suspension with adjustable spring constant. In Proceedings of the 19th International Conference on Solid-State Sensors, Actuators and Microsystems (TRANSDUCERS), Kaohsiung, Taiwan, 18–22 June 2017; pp. 934–937.

32. Poletkin, K.; Shalati, R.; Korvink, J.G.; Badilita, V. Pull-in Actuation in Micro-machined Hybrid Contactless Suspension. In Proceedings of the 17th International Conference on Micro and Nanotechnology for Power Generation and Energy Conversion Applications (Power MEMS 2017), Kanazawa, Japan, 14–17 November 2017; pp. 150–153.

33. Elata, D.; Bamberger, H. On the dynamic pull-in of electrostatic actuators with multiple degrees of freedom and multiple voltage sources. *J. Microelectromech. Syst.* **2006**, *15*, 131–140.

34. Poletkin, K.; Chernomorsky, A.; Shearwood, C.; Wallrabe, U. A qualitative analysis of designs of micromachined electromagnetic inductive contactless suspension. *Int. J. Mech. Sci.* **2014**, *82*, 110–121.

35. Poletkin, K.V.; Lu, Z.; Wallrabe, U.; Korvink, J.G.; Badilita, V. A qualitative technique to study stability and dynamics of micro-machined inductive contactless suspensions. In Proceedings of the 19th International Conference on Solid-State Sensors, Actuators and Microsystems (TRANSDUCERS), Kaohsiung, Taiwan, 18–22 June 2017; pp. 528–531.

36. Poletkin, K.; Lu, Z.; Wallrabe, U.; Korvink, J.G.; Badilita, V. Stable dynamics of micro-machined inductive contactless suspensions. *Int. J. Mech. Sci.* **2017**, *131–132*, 753–766.

37. Poletkin, K.V.; Chernomorsky, A.I.; Shearwood, C. A Proposal for Micromachined Dynamically Tuned Gyroscope, Based on Contactless Suspension. *IEEE Sens. J.* **2012**, *12*, 2164–2171.

38. Merkin, D.R. *Introduction to the Theory of Stability*; Springer Science & Business Media: Berlin, Germany, 2012; Volume 24.

39. Poletkin, K.; Chernomorsky, A.I.; Shearwood, C.; Wallrabe, U. An analytical model of micromachined electromagnetic inductive contactless suspension. In Proceedings of the ASME 2013 International Mechanical Engineering Congress & Exposition, San Diego, CA, USA, 15–21 November 2013; ASME: San Diego, CA, USA, 2013; p. V010T11A072.

40. Lu, Z.; Poletkin, K.; Wallrabe, U.; Badilita, V. Performance characterization of micromachined inductive suspensions based on 3D wirebonded microcoils. *Micromachines* **2014**, *5*, 1469–1484.

41. Rosa, E.; Grover, F. Formulas and tables for the calculation of mutual and self-inductance. *J. Wash. Acad. Sci.* **1911**, *1*, 14–16 .

42. Dwight, H.B. *Tables of Integrals and Other Mathematical Data*, 4th ed.; The MacMillan Company: New York, NY, USA, 1961.

![actuators logo] *actuators*

MDPI

Article

Learning Micromanipulation, Part 1: An Approach Based on Multidimensional Ability Inventories and Text Mining

Gaetano Biancucci [1], Giovanni Bonciani [1], Simona Fioravanti [1], Antonello Binni [1,*],
Franco Lucchese [2] and Apollonia Matrisciano [3]

[1] Department of Mechanical and Aerospace Engineering, University of Rome La Sapienza, 00184 Rome, Italy; biancucci.1541376@studenti.uniroma1.it (G.B.); bonciani.1430674@studenti.uniroma1.it (G.B.); fioravanti.1388838@studenti.uniroma1.it (S.F.)

[2] Department of Dynamic and Clinical Psychology, University of Rome La Sapienza, 00185 Rome, Italy; franco.lucchese@uniroma1.it

[3] Faculty of Industrial and Civil Engineering, University of Rome La Sapienza, 00184 Rome, Italy; lia.matrisciano@uniroma1.it

* Correspondence: antonello.binni@uniroma1.it; Tel.: +39-06-44585-680

Received: 3 August 2018; Accepted: 31 August 2018; Published: 3 September 2018

Abstract: In the last decades, an effort has been made to improve the efficiency of high-level and academic education players. Nowadays, students' preferences and habits are continuously evolving and so the educational institutions deal with important challenges, such as not losing attractiveness or preventing early abandonment during the programs. In many countries, some important universities are public, and so they receive national grants that are based on a variety of factors, on which the teaching efficiency has a great impact. This contribution presents a method to improve students commitment during traditional lessons and laboratory tests. The idea consists in planning some activities according to the students' learning preferences, which were studied by means of two different approaches. The first one was based on Gardner's multiple intelligence inventory, which is useful to highlight some peculiar characteristics of the students on the specific educational field. In the second method, direct interviews, voice recognition, and text mining were used to extract some interesting characteristics of the group of students who participated in the projects. The methods were applied in May 2018 to the students attending the course of Micro-Nano Sensors and Actuators for the postgraduate academic program dedicated to Industrial Nanotechnologies Engineering of the University of Rome La Sapienza. The present paper represents the first part of the investigation and it is dedicated essentially to the adopted methods. The second part of the work is presented in the companion paper dedicated to the presentation of the practical project that the students completed before the exam.

Keywords: teaching; learning; mocromanipulation; microactuation; multiple ability inventories

---

1. Introduction

Teaching methods and procedures at the high level of education and in Academia have been changing during the last years, whereas improving the efficiency of the educational process has become a crucial target for institutions [1,2]. Related studies show that improving the teaching–learning process has a certain impact on economic development [3], while the process itself is ineluctably affected by the mutual influences between technology and society [4] and by socio-emotional issues [5].

Furthermore, in many European countries, recent laws apply rewarding criteria to those institutions who support innovation in teaching and reduce the number of students abandoning or prolonging their courses.

In this paper, a recent experience of teaching–learning activities is presented, with the purpose of suggesting a way to improve students involvement and achievements. The adopted methods were applied in May 2018 to the students attending the course of Micro-Nano Sensors and Actuators for the Master degree in Industrial Nanotechnologies Engineering of the University of Rome La Sapienza. The present paper is dedicated essentially to the description of the teaching methods, whereas the consequent students' activities are presented in its companion paper [6].

An Approach Based on Multidimensional Ability Theoretical Framework

The basic idea for this investigation consisted in detecting the students' learning preferences and then tailoring some activities that could match the detected learning styles. This approach was expected to have the maximum effect in terms of achievements and satisfaction. The detection of the students' preference profile was approached by means of two different methods. The first one was inspired by the concept of Multidimensional Abilities, while the second one was based on text analysis.

Multidimensional Abilities have been investigated through well-known theoretical frameworks:

- Aptitude-Treatment Interaction [7];
- Conditions of Learning [8];
- Dual Coding Theory [9];
- Multiple Intelligences [10];
- Structure of Intellect [11]; and
- Triarchic Theory [12].

In this study, Gardner's approach was preferred because it has been applied in the past and gave some interesting results (see, for example, [13–16]). According to the original formulation of Multiple Intelligence inventory, the following intelligences are considered:

- Intrapersonal;
- Bodily-Kinesthetic;
- Interpersonal;
- Logical-Mathematical;
- Spatial;
- Verbal-Linguistic; and
- Musical.

Nowadays, Gardner's Multiple Intelligence is very popular and its principles have been described in many reference books and websites. In the present investigation, this inventory was applied to the specific context of learning in a technological education field, micromanipulation, in a high-level academic course. Consequently, the investigation must be considered as restricted to a rather superficial layer of one student's individual personality. The results do not concern the individual's general characteristics, but rather involve the individual's characteristics in the context of the specific environment, namely, a course in Industrial Nanotechnologies Engineering. To better explain this concept, let us review how the multiple intelligences have been applied in the context of learning in the academic environment.

Intrapersonal intelligence was not measured as an individual's general characteristic, but as a student's attitude of thinking and studying alone, without external interactions.

Bodily-Kinesthetic intelligence was associated with the use of practical activities that involve manual handling of models or instruments in a lab. This attitude was also associated with the use of physical support for studying and understanding.

Interpersonal intelligence was here related to some students attitude to gain benefit from social activities, such as studying in the company of other people and participating or promoting small social events.

Logical-Mathematical intelligence was associated with the students' preference for using logic and mathematical formulation to understand new models and concepts.

Spatial intelligence was related to a student preference for using graphical tools, such as pictures, diagrams, block charts, or spatial models.

Verbal-Linguistic intelligence was related to the students' preference of learning by reading or hearing some text.

Finally, Musical intelligence was measured as in standard tests but was not related to any dimension for the sake of the present investigation.

2. Material and Methods

A questionnaire was submitted to a limited group of students to measure their singular preference profiles based on Gardner's inventory. The adopted test has the advantage that it has been used before [16] on a reference group and on some groups of students enrolled in the courses of Engineering at Sapienza University. Therefore, averages and standard variations on raw scores were already available for the adopted dimensions. This circumstance was quite important because the number of participating students was rather limited ($n = 6$).

2.1. Correlation Analysis among Dimensions

As for the previous investigations [14–16], standard statistical analysis was applied to the raw scores obtained from the surveys. Then, standard z-scores were calculated by an elementary routine. z-scores are a standardization of the raw data which maps scores in such a way that the standardized population has a null mean and a unit standard deviation (see Figure 1).

Figure 1. Individuals standard z-scores for the adopted seven dimensions.

The adoption of the z-scores was useful because the calculation of the correlation matrix becomes quite easy. Table 1 shows the classical correlation table for the seven dimensions under analysis. Since only six individuals participated in the survey, the obtained results are not significant for possible generalization and so the curious positive or negative correlations have a value restricted to the analyzed group.

Table 1. Correlation table for the group of participating students ($n = 6$).

	Verbal	Logical	Musical	Kinesthetic	Spatial	Interpersonal	Intrapersonal
Verbal	1.00	−0.32	0.35	0.43	0.47	−0.06	−0.59
Logical		1.00	−0.57	−0.71	−0.57	0.01	0.03
Musical			1.00	0.39	−0.01	0.03	−0.63
Kinesthetic				1.00	0.81	0.64	−0.26
Spatial					1.00	0.47	0.15
Interpersonal						1.00	−0.14
Intrapersonal							1.00

Figure 2 shows the scatter diagrams which illustrate an example of both positive and negative correlations among two pairs of dimensions. As mentioned above, it is not possible to draw a general conclusion that any student who has a kinestetic preference must have necessarily a great spatial-visual attitude and a small Logical-Mathematical preference. This is only a trend that appeared in a small class.

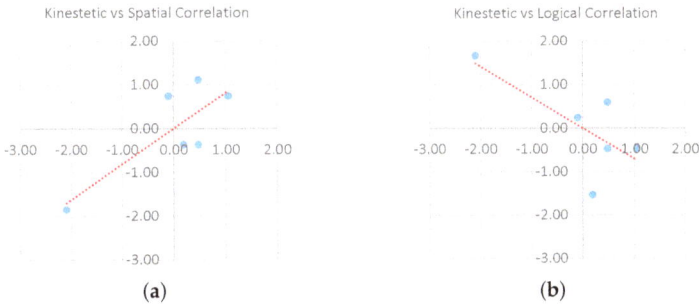

(a)   (b)

Figure 2. Scatter diagram for two pairs of dimensions: (**a**) Spatial-Visual vs. Bodily-Kinesthetic positive correlation; and (**b**) Spatial-Visual vs. Logical-Mathematical negative correlation.

2.2. Characteristics of the Class Group

The present investigation attempted to ascertain whether the small class was characterized by peculiar characteristics. Of course, this analysis must be done by comparing the dimensions averages of the class with those obtained in a control group. Fortunately, the same questionnaire had been previously submitted to a much larger control group during an earlier investigation [16], and so these results were used for the present analysis. For each of the seven dimensions, the pairs of averages were compared by pointing out those with a minimum value for the significance levels. Hence, the null hypothesis $\mathcal{H}_0$ was set as the assumption that the class and control groups were extracted from the same population.

Weak ($0.01 < \alpha \leq 0.1$) or strong ($0.001 \leq \alpha \leq 0.01$) probability levels were adopted together with Student t-test and the degrees of freedom $n_A = 6$ and $n_B = 56$ corresponding to class and control group populations, respectively. Some significant differences are reported in Table 2.

Table 2. Significant differences between raw score dimensions of the class and control group populations.

	Verbal	Logical	Musical	Kinesthetic	Spatial	Interpersonal	Intrapersonal
	lower	greater	lower	greater	greater	greater	lower
p	>0.1	0.025	>0.1	0.1	0.001	0.001	>0.1
	no	weak	no	weak	strong	strong	no

- Verbal (or Linguistic) Intelligence: According to previous investigations, students enrolled in high-level Engineering courses present verbal capabilities usually higher than their peers who do not frequent Academic Institutions. Our class students present the same characteristic as the control group students, and therefore no significant difference was found for this dimension.

- Logical-Mathematical Intelligence: Our students showed a preference for logical and mathematical formalism weakly greater than the students belonging to the control group. This is quite reasonable for students enrolled in the last year of a master course in Nanotechnology.
- Musical Intelligence: No significant difference was detected for this dimension.
- Kinesthetic Intelligence: This dimension is characterized by a weakly significant difference in the average scores for the two groups; actually, engineering students are quite handy and so our students simply showed an attitude a little more pronounced than the control group.
- Spatial-Visual Intelligence: According to the significance analysis, it is quite sure that our students prefer an approach based on practical observation of planar schemes and 3D objects manipulation more than peers students enrolled in other courses; this dimension suggests the use of real objects and the practical manipulation of real models as an effective teaching method for this class.
- Interpersonal Intelligence: The clear preference in the class of an approach based on interpersonal intelligence is a straightforward suggestion to adopt interactive lessons with a considerable amount of teamwork.
- Intrapersonal Intelligence: No significant difference with respect to the control group was detected for this dimension.

3. Text Mining for the Identification of Patterns

Before the starting of the course, some interviews were arranged with the students. The transcripts of the oral interviews were automatically obtained by means of available software with the simple aid of a smartphone.

The following questions were submitted to the students before the beginning of the course.

1. Which is your personal motivation for you to study Nanotechnology?
2. What do you think about the activities in the lab?
3. What is your opinion about manual activities?
4. Do you think they can help your understanding of the lessons?

The following questions were submitted to the same students during of the course.

5. Did you change your opinion about the activities in the lab?
6. How much do you feel adequate for this experience in the lab?
7. What are your personal features that you think they are the most suitable for this experience?

These questions were used as cognitive stimulus to trigger the explication of concepts from the interviewees. The questions had the function of making some concepts more visible than in the case of a free speech. The advantage of this technique consists in the possibility of handling quantitatively some parameters of interest.

Then, the answers were filtered to remove from text some insignificant part of the spoken language and to obtain a polished text (conjunctions, prepositions, articles, and so on). The latter was then analyzed using text mining classification and frequency analysis of words occurrences. The graphic technique of words cloud was applied to extract key concepts and the most used words, and a graphic composition of such collection is reported in Figure 3, where dimensions and colors are related to the occurrence frequency. This wordcloud representation is very interesting to gain a picture of the tags that were most raised during the conversations with students.

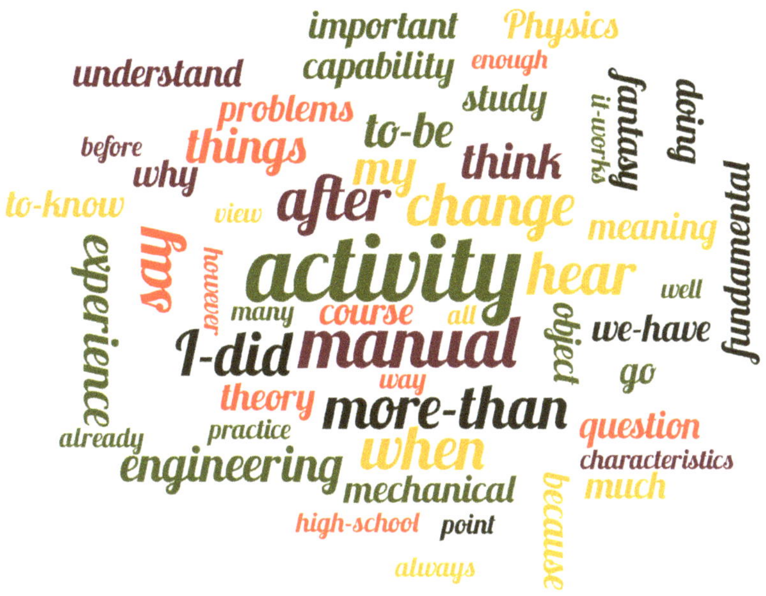

Figure 3. Most frequently mentioned words during interviews.

The following are some examples of frequently spoken words during the interview (see also Figure 3), and some reasonable interpretations.

- activity, experience, things, practice, hand, it-works, manual, and doing: These words are quite central during the interview; they are related to the students direct involvement to the lab activities and their willingness to learn by means of direct experience rather than by means of the classical textbook study.
- view and hear are also related to senses and so to the experience.
- more than, when, after, before, change, and go: These words are clearly related to time, development and progress and therefore they show the students willingness to improve, compete and face challenges.
- engineering, mechanical, Physics, high school, and study: These are clearly related to the context of the course.
- think, understand, meaning, fundamental, and important: These are related to cognitive and learning process.
- my and I-did: These are probably related to the students positive self promotion, and to the concept of making valuable their own experience.
- fantasy, finally and curiously, appears as a possible sign of passion for the topic and a windows on creativity in design.

The same text mining technique was applied to the course syllabus and the most cited words are presented in a graphical word cloud (Figure 4). A comparison of the two clouds suggests that the main differences consist of terms that represent action and personal experience. Once again, the comparison shows how this class students of were sensible to the manual and practical activities while learning a topic which implies, intrinsically, some theoretical concepts. In fact, micro and nano actuation is a topic which requires multiple (and rather opposite) abilities such as theoretical and practical skills. This is a possible reason for the success of the practical projects reported in the second part of this investigation [6].

The graphics in Figures 3 and 4 were generated by means of Wordsalad © free app [17].

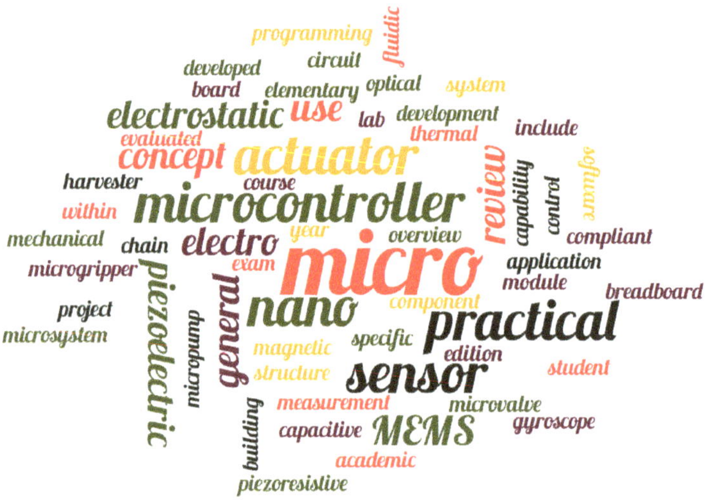

Figure 4. Most frequently mentioned words in the course syllabus.

4. Results and Discussion

Considering the significance analysis described in the previous sections, our class presents three significant differences with respect to the control group: Spatial-Visual, Interpersonal and Kinesthetic Intelligences. The results obtained from the text analysis also showed an evident inclination of the students to the activities and practical work and therefore they confirmed the results obtained by using Gardner's inventory. Therefore, considering that the Course basic material is based on classical text books [18,19] and on some recent papers [20–31], i.e., written material, the peculiar above-mentioned results gave rise to the idea of encouraging our students to work in team (Interpersonal) and make use of handmade (Kinesthetic) models (Spatial-Visual).

Since this is not possible for nano or micro size devices, the idea of using properly up-scaled models arose. By assembling the indications from the Multiple Intelligences method and from the text analysis together, it was clear that some practical activities would have given a great boost to teaching and therefore some projects were designed for the specific group of students.

These projects have the following characteristics.

- From the course program, a micro or nano device was selected as the case study.
- A series of activities is planned for the selected device, according to the common operational capability required for the real micro or nano system under study.
- A macroscale model is obtained for the micro or nano device by using 3D printing low-cost techniques.
- The operational features are translated from the micro- (or nano-) to the macroscale, possibly by changing some principles of operation (this allows the students to understand some fundamental concept of scaling in fabrication and actuation).
- During these activities tinkering is encouraged for the construction of the system at the macroscale.
- Finally, a system is created at the macroscale based on the micro- or nanosystem, and the differences between the two systems are material for the exam.

Some of these projects are described in detail in the companion paper [6].

Considering the excellent scores achieved by the students during the exams and their great satisfaction, it is possible to conclude that the presented teaching–learning experience had quite a positive impact on the course. More generally, results show that, thanks to the adopted methods, the quality of teaching/learning can be improved also in the context of Nanotechnology and Micro Actuation.

Author Contributions: Data Curation, Writing—Original Draft Preparation, all the Authors; Conceptualization, Methodology, and Supervision, F.L. and A.M.; Software and Formal Analysis, G.B. (Gaetano Biancucci), G.B. (Giovanni Bonciani) and S.F.; Resources, Writing—Review & Editing, A.B.

Funding: This research received no external funding.

Conflicts of Interest: The authors declare no conflict of interest.

References

1. Renaud, R.; Murray, H. The validity of higher-order questions as a process indicator of educational quality. *Res. High. Educ.* **2007**, *48*, 319–351. [CrossRef]
2. Mechtenberg, L.; Strausz, R. The Bologna process: How student mobility affects multi-cultural skills and educational quality. *Int. Tax Public Financ.* **2008**, *15*, 109–130. [CrossRef]
3. Castelló-Climent, A.; Hidalgo-Cabrillana, A. The role of educational quality and quantity in the process of economic development. *Econ. Educ. Rev.* **2012**, *31*, 391–409. [CrossRef]
4. Belfiore, N.P.; Di Benedetto, M.; Matrisciano, A. The mutual influences between technology and society: A contribution from the Mechanical Engineering Diploma Universitario of the University of Rome La Sapienza. In Proceedings of the International Symposium on Technology and Society, Rome, Italy, 8 September 2000; pp. 1–5.
5. García Palomo, M.J.; Sánchez, J.L.R.; Herrera, S.S.; Briegas, J.J.; Lucchese, F. Influence of a socio-emotional learning program in the emotional intelligence as applied skill [Influencia de un programa de aprendizaje socio-emocional sobre la inteligencia emocional autopercibida]. *Confin. Cephalalgica* **2018**, *28*, 16–24. (In Spanish)
6. Bonciani, G.; Biancucci, G.; Fioravanti, S.; Valiyev, V.; Binni, A. Learning micromanipulation, part 2: Term projects in practice. *Actuators* **2018**, submitted.
7. Snow, R. Aptitude-Treatment Interaction as a Framework for Research on Individual Differences in Psychotherapy. *J. Consult. Clin. Psychol.* **1991**, *59*, 205–216. [CrossRef] [PubMed]
8. Richey, R.C. (Ed.) *The Legacy of Robert M. Gagne*; Syracuse University; ERIC Clearinghouse on Information & Technology: New York, NY, USA, 2000.
9. Clark, J.M.; Paivio, A. Dual coding theory and education. *Educ. Psychol. Rev.* **1991**, *3*, 149–210. [CrossRef]
10. Gardner, H. *Frames of Mind: The Theory of Multiple Intelligences*; BasicBooks: New York, NY, USA, 1993.
11. Guilford, J.P. *The Nature of Human Intelligence*; McGraw-Hill: New York, NY, USA, 1967.
12. Sternberg, R.J. *Intelligence, Information Processing, and Analogical Reasoning: The Componential Analysis of Human Abilities*; Lawrence Erlbaum Associates: Mahwah, NJ, USA, 1977.
13. Belfiore, N.P.; Rudas, I.; Matrisciano, A. Simulation of verbal and mathematical learning by means of simple neural networks. In Proceedings of the 9th International Conference on Information Technology Based Higher Education and Training, Chicago, IL, USA, 29 June–2 July 2010; pp. 52–59.
14. Matrisciano, A.; Belfiore, N.P. An investigation on cognitive styles and multiple intelligences model based learning preferences in a group of students in engineering. In Proceedings of the 9th International Conference on Information Technology Based Higher Education and Training, Chicago, IL, USA, 29 June–2 July 2010; pp. 60–66.
15. Matrisciano, A.; Deplano, V.; Belfiore, N.P. Analysis of a teaching and learning method supported by open source codes and web activities. In Proceedings of the International Conference on Information Technology Based Higher Education and Training, Istanbul, Turkey, 21–23 June 2012.
16. Micangeli, A.; Naso, V.; Michelangeli, E.; Matrisciano, A.; Farioli, F.; Belfiore, N.P. Attitudes toward sustainability and green economy issues related to some students learning their characteristics: A preliminary study. *Sustainability* **2014**, *6*, 3484–3503. [CrossRef]
17. Spagnolini, L.; Cerutti, A.; Pagliara, G. Wordsalad: Your Word Clouds Redefined. Available online: wordsaladapp.com (accessed on 1 August 2018).

18. Bhushan, B. (Ed.) *Springer Handbook of Nanotechnology*; Springer Handbooks: Berlin/Heidelberg, Germany, 2017.

19. Gad-el-Hak, M. *The MEMS Handbook—3 Volume Set*, 2nd ed.; Mechanical and Aerospace Engineering Series; CRC Press: Boca Raton, FL, USA, 2005.

20. Balucani, M.; Belfiore, N.P.; Crescenzi, R.; Verotti, M. The development of a MEMS/NEMS-based 3 D.O.F. compliant micro robot. *Int. J. Mech. Control* **2011**, *12*, 3–10.

21. Belfiore, N.P.; Simeone, P. Inverse kinetostatic analysis of compliant four-bar linkages. *Mech. Mach. Theory* **2013**, *69*, 350–372. [CrossRef]

22. Verotti, M.; Crescenzi, R.; Balucani, M.; Belfiore, N.P. MEMS-based conjugate surfaces flexure hinge. *J. Mech. Des. Trans. ASME* **2015**, *137*. [CrossRef]

23. Belfiore, N.P.; Broggiato, G.; Verotti, M.; Balucani, M.; Crescenzi, R.; Bagolini, A.; Bellutti, P.; Boscardin, M. Simulation and construction of a mems CSFH based microgripper. *Int. J. Mech. Control* **2015**, *16*, 21–30.

24. Cecchi, R.; Verotti, M.; Capata, R.; Dochshanov, A.; Broggiato, G.; Crescenzi, R.; Balucani, M.; Natali, S.; Razzano, G.; Lucchese, F.; et al. Development of micro-grippers for tissue and cell manipulation with direct morphological comparison. *Micromachines* **2015**, *6*, 1710–1728. [CrossRef]

25. Di Giamberardino, P.; Bagolini, A.; Bellutti, P.; Rudas, I.; Verotti, M.; Botta, F.; Belfiore, N. New MEMS tweezers for the viscoelastic characterization of soft materials at the microscale. *Micromachines* **2017**, *9*, 15. [CrossRef]

26. Bagolini, A.; Ronchin, S.; Bellutti, P.; Chistè, M.; Verotti, M.; Belfiore, N.P. Fabrication of Novel MEMS Microgrippers by Deep Reactive Ion Etching with Metal Hard Mask. *J. Microelectromech. Syst.* **2017**, *26*, 926–934. [CrossRef]

27. Dochshanov, A.; Verotti, M.; Belfiore, N. A Comprehensive Survey on Microgrippers Design: Operational Strategy. *J. Mech. Des. Trans. ASME* **2017**, *139*, 070801. [CrossRef]

28. Verotti, M.; Dochshanov, A.; Belfiore, N.P. A Comprehensive Survey on Microgrippers Design: Mechanical Structure. *J. Mech. Des. Trans. ASME* **2017**, *139*, 060801. [CrossRef]

29. Verotti, M.; Dochshanov, A.; Belfiore, N.P. Compliance Synthesis of CSFH MEMS-Based Microgrippers. *J. Mech. Des. Trans. ASME* **2017**, *139*, 022301. [CrossRef]

30. Potrich, C.; Lunelli, L.; Bagolini, A.; Bellutti, P.; Pederzolli, C.; Verotti, M.; Belfiore, N.P. Innovative silicon microgrippers for biomedical applications: Design, mechanical simulation and evaluation of protein fouling. *Actuators* **2018**, *7*, 12. [CrossRef]

31. Crescenzi, R.; Balucani, M.; Belfiore, N.P. Operational characterization of CSFH MEMS technology based hinges. *J. Micromech. Microeng.* **2018**, *28*. [CrossRef]

actuators

MDPI

Article

Learning Micromanipulation, Part 2: Term Projects in Practice

Giovanni Bonciani, Gaetano Biancucci, Simona Fioravanti, Vagif Valiyev and Antonello Binni *

Department of Mechanical and Aerospace Engineering, University of Rome La Sapienza, 00184 Rome, Italy;
bonciani.1430674@studenti.uniroma1.it (G.B.); biancucci.1541376@studenti.uniroma1.it (G.B.);
fioravanti.1388838@studenti.uniroma1.it (S.F.); valiyev.1773704@studenti.uniroma1.it (V.V.)
*   Correspondence: antonello.binni@uniroma1.it; Tel.: +39-06-44585-287

Received: 3 August 2018; Accepted: 31 August 2018; Published: 3 September 2018

Abstract: This paper describes the activities that have been necessary to design, fabricate, control and test some low-cost test stands independently developed by the students enrolled in the course of Micro-Nano sensors and actuators for the postgraduate course in Industrial Nanotechnologies Engineering of the University of Rome La Sapienza. The construction and use of these test stands are an essential part of teaching and learning methods whose theoretical bases have been presented in the companion paper (Part 1). Each test stand is composed of a compliant structure and a control system, which consists of a programmable control micro-card equipped with sensors and actuators. The compliant structure consists of a compliant mechanism whose geometry is achieved by scaling some previously developed silicon micromanipuators and microactuators up to the macroscale by a factor of 20. This macroscale model offered a kinesthetic tool to improve the understanding of the original microsystems and their working principles. The original silicon micromechanisms have been previously presented in the literature by the research group after design and deep reactive-ion etching (DRIE) microfabrication. Scaling from micro to macro size was quite easy because the original DRIE masks were bestowed to the students in the form of CAD files. The samples at the macroscale have been fabricated by means of recently available low-cost 3D printers after some necessary modifications of the mask geometry. The purpose of the whole work (Parts 1 and 2) was the improvement of the efficiency of an educational process in the field of microsystem science. By combining the two companion papers, concerning, respectively, the theoretical basis of the teaching methods and the students' achievements, it is possible to conclude that, in a given class, there may be some preferred activities that are more efficient than others in terms of advancements and satisfaction.

Keywords: teaching; learning; mocromanipulation; microactuation; multiple ability inventories

1. Introduction

This paper presents some activities that have been carried out during the class of Micro-Nano sensors and actuators for the postgraduate academic program dedicated to Industrial Nanotechnologies Engineering of the University of Rome La Sapienza. The first part of the investigation [1] presented two didactical methods, the first one being based on Gardner's multiple intelligence inventory and the second one based on text analysis. Multiple intelligences have been used to reveal a characteristic profile of each student and some significant characteristics of the whole group. Secondly, text analysis has been applied to the transcript of some interviews with the students and also to the course syllabus, as a reference text for comparison. The two methods showed that this particular class had a strong inclination to the practical and interpersonal work. Based on those results, the teaching staff decided to stress the importance of term projects because they allow students to get easily familiar with the micro world through their preferred kinesthetic approach and motivational feelings. The present paper describes the projects that have been carried out by the students their self during the semester. This preferences–based teaching–learning method has the advantage of increasing the course attractiveness, participation and efficiency in terms of students achievements.

Since the students participate in the above-mentioned course during their last year of the Master degree, the teaching material must be quite upgraded and also at the scientific literature level, because micro and nano technology is rather rapidly changing. Other than studying some portions of classic text books on micro and nano technology [2,3], more specific readings were suggested, related to the previous experience achieved by the teaching staff and concerning compliant 3 D.O.F. microrobots [4–7], micromechanisms [8], micro hinges [9–12] and microgrippers [13–18]. However, this material is essentially written text.

As explained in the companion paper [1], the verbal approach is not the only way students learn. For the class group involved in the present investigation, some kinesthetic activities have been encouraged during laboratory meeting. The idea of increasing students–object tactile interactions comes from the fact that course subject consists of micro and nano objects, which are even hardly visible. Therefore, there are few ways of presenting them: showing the masks at a certain magnification, producing SEM or optical images or short animations of the objects, or through verbal description. In all these cases, there is scarce direct or manual interaction. Nevertheless, it has been shown that kinesthetic exercises may activate human original corporeal–kinetic apprenticeship [19].

During lab activities, the students have been provided with a certain number of silicon prototypes for observation and analysis. However, each sample consists in a device that is embedded within a 2 square millimeter, overall size, with some specific parts, such as the comb drive fingers, which have two-micron width. This makes direct interaction practically impossible. Bringing the microsystem to the macroscale is useful not only for simulation but also for teaching purposes. In fact, the students become able to develop their own simplified systems that can be also operated at the macroscale by means of inexpensive micro-cards. These activities can be also useful to stimulate a certain capability of working in team. Finally, thanks to the development of new products and to some new methods based on tinkering [20], the overall cost of these lab activities can be extremely limited, if compared to the usual costs of equipment used in the common nanotechnology lab, while the students involvement remains very high. The term projects described in the next section have been developed by the students under the surveillance and advice of the teaching staff.

2. Material and Methods

2.1. Fabrication of the Mechanical Structure

The widespread success of additive manufacturing made 3D printers available at low costs. Many students and practitioners have been able to assemble their own 3D printers by using limited resources and this circumstance made these devices very popular among students. For this reason, the idea of using a 3D printer for scaling up a microsystem, although not very original, has been accepted with a certain enthusiasm by the students.

The students were first provided with the original mask CAD files that have been used to fabricate the real microsystems. Then, they modified the 2D geometry to adjust the beams or the fingers minimum width to the maximum resolution of the adopted printer. The 3D printer was also used to fabricate the housing for the compliant mechanism, the motors, the micro-cards and all the necessary mechanical parts, in such a way that the whole system would be correctly assembled on the frame.

2.2. Double Rotary Link Microgripper

The first project consisted in the fabrication and the experimental operation at the macroscale of a simple double rotary link gripper. The students were provided with the original mask CAD files for the DRIE (Deep Reactive Ion Etching) process (see Ref. [14] for details). The actual design was developed by means of AutoCAD© (Version 2015, Autodesk, San Rafael, CA, USA), and so the starting point, for the students, was the mask depicted in Figure 1a. This design was scaled as well as modified to fit the limits of the 3D printer accuracy to avoid unfeasible widths of fingers, gaps and curved beams. The truss structure embedded in the original links was also simplified, in order not to make critical the process of 3D printing. Finally, the fingers of the mobile comb drives were eliminated for the sake of simplicity.

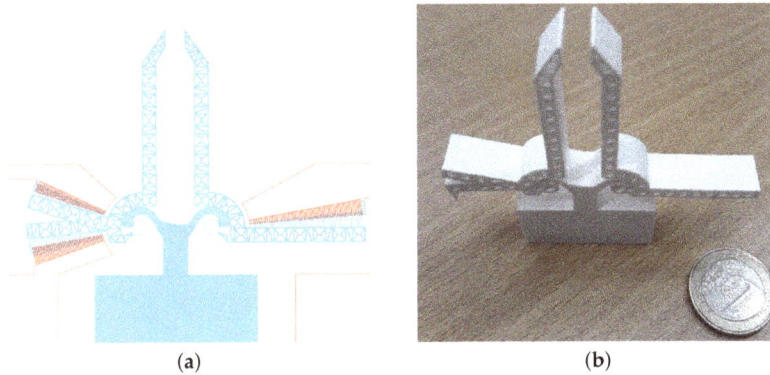

(a) (b)

Figure 1. The original mask for a compliant rotary microgripper (**a**); and its prototype at the macroscale (**b**) (first project).

The AutoCAD *dwg* file was translated into the classical *stl* file and then provided to the 3D printer, which yielded the layout represented in Figure 1b.

The macroscale compliant mechanisms were subject to preliminary tests which gave encouraging results in terms of mechanical properties (compliance and robustness). Then, the experimental layout was set for operational tests. Since the comb drives were not feasible at the macroscale, their action was simulated by means of cables attached to the end of the comb drive mobile wings and operated by a motor. A simplified scheme is reported in Figure 2, where the two wings, the two cables and the two motor arms are represented, respectively, by lines AB and DE, BC and EF, and CO and FO. When the cables are under tension, the groups ABCO and DEFO can be considered as four-bar linkages, which makes it very easy to perform the kinematic analysis of the system. In the case under study, the forward run was achieved by applying a counterclockwise torque, while the return run was obtained simply by releasing the above mentioned torque, by virtue of the elastic energy stored in the flexure revolute joints A and D.

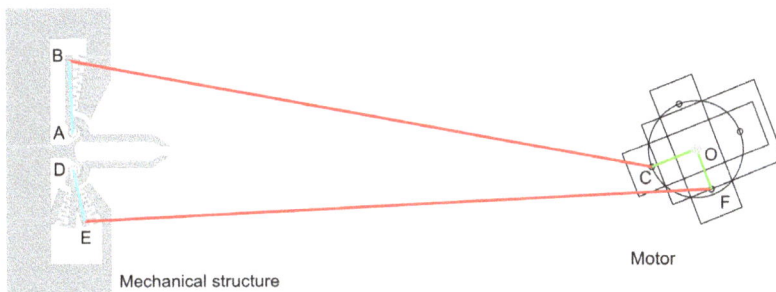

Figure 2. Actuation of the compliant mechanisms at the macroscale (first project).

The experimental layout was completed using the same 3D printer for building the mechanical anchors and frames. The printing parameters were set by using Ultimaker Cura software©

(Version 3.2.1), and the G-code file was prepared. All components were re-printed several times and then assembled to the base. Shaking torques transmitted to the base were estimated and a proper frame was set on the motors to allow them to work adequately. Electronic circuits and control card were mounted to the same base, as a reminder that the real microsystem would be arranged on a single chip, by using ASIC (Application Specific Integrated Circuit). The arrangement is illustrated in Figure 3.

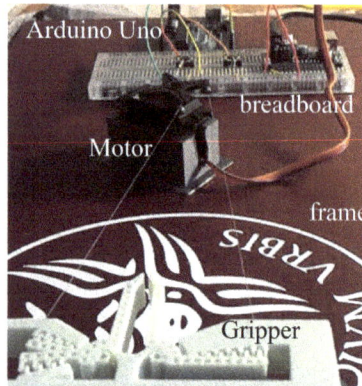

Figure 3. A picture of the experimental layout (first project).

The actual circuit was developed using Fritzing© (Version 0.9.3), and the wiring scheme was realized by using a common breadboard, which includes also connection to Arduino UNO© (REV3). As is usual for this particular microcontroller, a simple Arduino IDE© code (Version 1.8.5), was developed, tested and uploaded on the microcard, in such a way that the whole system no longer needed direct connection to the computer. Forces were transmitted to the compliant mechanisms through nylon cables. In the first configuration, the closing motion was obtained by exerting a force on the jaws levers, while the opening return motion is achieved by releasing these forces and letting the flexure elasticity to drive back the jaws to their original position.

2.3. Bidirectional Double Four-Bar Linkage Microgripper

In the second project, a bidirectional microgripper, shown in Figure 4a, was scaled up to the macroscale sample, represented in Figure 4b, by following the same method of design, printing and actuation as described in the previous paragraph. The finger cross section was scaled from $2\ \mu m \times 40\ \mu m$ to $0.6\ mm \times 1\ cm$, while the microgripper overall dimensions are $10.8\ cm \times 10.3\ cm \times 1\ cm$. The macroscale prototype was obtained in Polylactic Acid, while actuation and control was obtained by Arduino microcontroller integrated with breadboard, circuits and two rotary actuators. The comb drives were mechanically controlled by means of cables that are attached in such a way that open and close maneuvers are allowed.

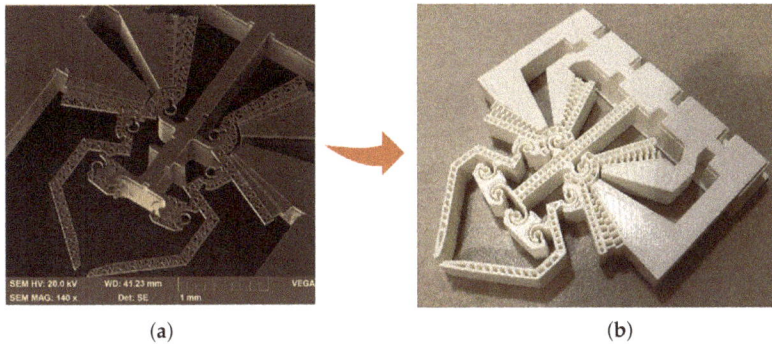

Figure 4. Comparison of the micro (see Ref. [14]) (**a**) and macro (**b**) grippers (second project).

The main problem with design consisted of the limited resolution of the 3D Delta printer. With the available model, the resolution was limited to 0.4 mm, thus some sections of the gripper had to be redesigned to cope with this limit. In fact, after linear scaling from the micro to the macro scale, there were still some parts (fingers and curved beams) that were too small for that resolution. Furthermore, it was also not possible to use the above-mentioned minimum size to not slow down the printing process too much, and so the minimum thickness had to be increased to 0.6 mm for both finger and gap between them. AutoCAD 2018 student version was used for mask scaling and redesigning (estimated time, non-expert user, 30 h for one person).

Ultimaker Cura© (version 3.2.1), was used for managing the printing process and changing some printer properties and maximize the likelihood of success. The main adopted changes in the printer properties were: layer height (0.1 mm), wall thickness (0.4 mm), printing temperature of the nozzle (220 °C), build plate temperature (30 °C) and print speed (35 mm/s), while cooling was enabled and the build plate adhesion type set to Skirt. After the printing stage, some surface grinding operations were performed to polish the surfaces (estimated time, non-expert users, 20 h for one person).

Similar to the first project, the system at the macroscale was made independent of the computer by using Arduino UNO© capability of uploading IDE codes ready to work with the system. The actuation scheme is represented in Figure 5. In this case, there are two motors which pull the comb drive wings to open the jaws. With reference to Figure 5, cable FE pulls body CDE to rotate around flexure center D, in such a way that the coupler CB drives the attached jaw toward the closure position.

The motors were controlled by means of a circuit that is similar to that described in the previous paragraph, being composed of a breadboard, some simple electronic components and a microcontroller. A picture of the experimental set is reported in Figure 6.

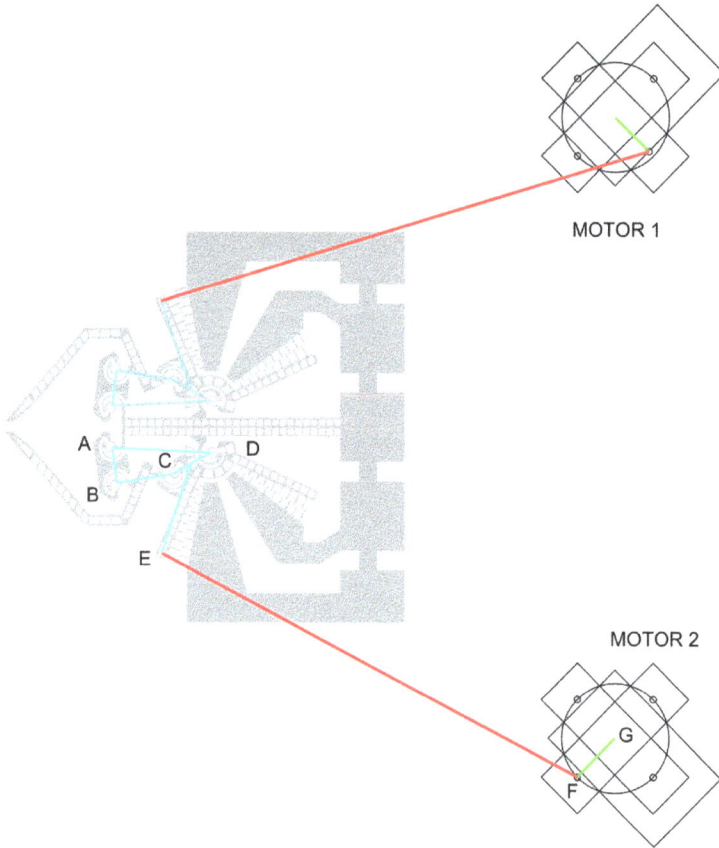

Figure 5. Actuation of the compliant mechanisms at the macroscale (second project).

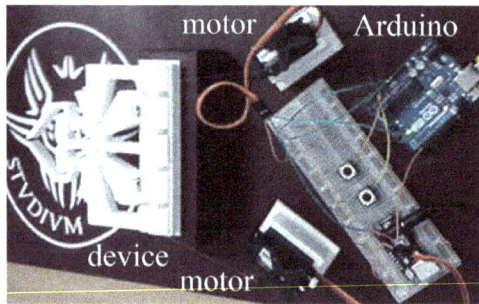

Figure 6. A picture of the experimental layout (second project).

3. Results and Discussion

The projects described in the previous section are useful mainly for educational purposes. In fact, the final layout is very similar to a chip that could be used for the real microsystems. The experimental scaling was also important for the student to understand that actuation is strongly related to size. The macroscale grippers were actuated using classical electric motors, while the microgrippers were actuated by means of comb drives. Nevertheless, the projects, as a whole, were also interesting for research purposes, because they were useful to point out all the necessary operations and components.

The attempt to build a system uniformly scaled from micro- to macroscale was used to appreciate some basic and very well known concepts of actuation at the microscale. In fact, the students were encouraged to build a comb drive at the macroscale similar to the one depicted in Figure 7a,b, which they did, of course, with no success. Therefore, the real actuation at the microscale was attempted by arming the wings of two symmetric magnetic circuits (see Figure 8a) with conductive media and then providing magnetic fields to the base of the system to achieve motion (see Figure 8b).

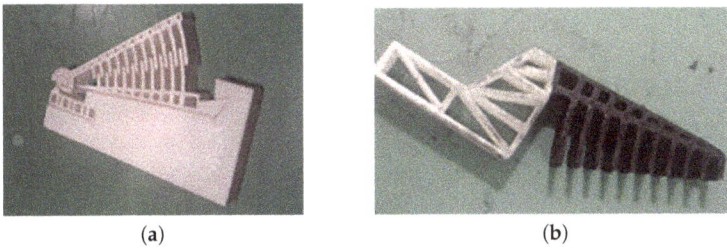

(a)   (b)

Figure 7. Unsuccessful comb drive at the macroscale: (**a**) bulk; and (**b**) covered by conductive paste.

(a)   (b)

Figure 8. Magnetic actuation at the macroscale: (**a**) bulk *stl* file (left and right wings); and (**b**) motion sequence of the armed wings.

These two simple experiments stimulated the discussion concerning the effects of scaling on actuation. In particular, the students were encouraged to figure out the consequences of a uniform reduction of the geometry by a factor of 10. As known, such reduction gives rise to a reduction of the electrostatic force by 100 times, while the corresponding reduction of the curved beam rotary stiffness is 1000. Considering that the electromagnetic force is reduced 10,000 times, it is clear how electrostatic actuators perform much better than the electromagnetic at the microscale.

The benefit of these activities was not limited to the specific course, but it was also extended to more general skills, such as working in a team, rapid prototyping, actuation, control and mechanical design. In particular, the students had to solve specific problems that are of a certain interest today.

The first problem was the optimal setting of the 3D printing system parameters, by taking into account some specific properties such as the nozzle section (0.4 mm) and the material (Polylactic acid, PLA). Then, the operational parameters had to be considered, such as the printing temperature (210–215 °C), and speed. Furthermore, different actuation strategies were analyzed and compared. After exclusion of comb drives, for losing efficacy after dimensional scaling, the servomotors power supply was originally arranged directly from one Arduino 5 V pin. This solution gave rise to several problems, such as torque ripples, unstable position and overheating due to excessive current. Therefore, the classical external power supply scheme with grounding GND pin has been arranged for both the motors, while Arduino provided control. Finally, safety principles have been also implemented during the activities, especially by means of personal protective equipment (PPE).

4. Conclusions

This paper has shown a way to improve the teaching activities in the field of microsystems and micro-manipulation. The investigation has proven how the complex teaching–learning process may be improved by tuning lab activities to the specific students' characteristics, expressed in terms of multiple intelligence inventory, interest, and motivation. The investigation has also proven that the use of low-cost, tinkering approach boosts the student involvement up to a level of interdisciplinary knowledge which was good enough for them to build a system complete with mechanical structure, transmission, actuation, and control. Students declared their satisfaction and excitement during the project, while the final exams showed that their understanding of phenomena at the microscale improved much after the tactile activities at the macroscale. The authors hope that the two papers will be somehow useful to professors for employing new methods of interactive and dedicated teaching for their laboratory courses.

Author Contributions: Conceptualization, all the authors; Data curation, Software, G.B. (Giovanni Bonciani), G.B. (Gaetano Biancucci), S.F. and V.V.; Resources, Writing—review & editing, A.B.

Funding: This research received no external funding.

Conflicts of Interest: The authors declare no conflicts of interest.

References

1. Biancucci, G.; Bonciani, G.; Fiore, S.; Binni, A.; Lucchese, F.; Matrisciano, A. Learning micromanipulation, part 1: An approach based on multidimensional ability inventories and text mining. *Actuators* **2018**, submitted.
2. Gad-el Hak, M. (Ed.) *The MEMS Handbook*, 2nd ed.; Mechanical and Aerospace Engineering Series; CRC Press, Tayor and Francis Group: Boca Raton, FL, USA, 2005.
3. Bhushan, B. (Ed.) *Springer Handbook of Nanotechnology*; Springer: Berlin/Heidelberg, Germany, 2017.
4. Balucani, M.; Belfiore, N.P.; Crescenzi, R.; Genua, M.; Verotti, M. Developing and modeling a plane 3 DOF compliant micromanipulator by means of a dedicated MBS code. In Proceedings of the Nanotech 2011: NSTI Nanotechnology Conference and Expo, Boston, MA, USA, 13–16 June 2011; Volume 2, pp. 659–662.
5. Balucani, M.; Belfiore, N.P.; Crescenzi, R.; Verotti, M. The development of a MEMS/NEMS-based 3 D.O.F. compliant micro robot. *Int. J. Mech. Control* **2011**, *12*, 3–10.
6. Belfiore, N.P.; Balucani, M.; Crescenzi, R.; Verotti, M. Performance analysis of compliant mems parallel robots through pseudo-rigid-body model synthesis. In Proceedings of the ASME 2012 11th Biennial Conference on Engineering Systems Design and Analysis, Nantes, France, 2–4 July 2012; Volume 3, pp. 329–334.
7. Belfiore, N.P.; Emamimeibodi, M.; Verotti, M.; Crescenzi, R.; Balucani, M.; Nenzi, P. Kinetostatic optimization of a MEMS-based compliant 3 DOF plane parallel platform. In Proceedings of the ICCC 2013—IEEE 9th International Conference on Computational Cybernetics, Tihany, Hungary, 8–10 July 2013; pp. 261–266.
8. Belfiore, N.P.; Simeone, P. Inverse kinetostatic analysis of compliant four-bar linkages. *Mech. Mach. Theory* **2013**, *69*, 350–372. [CrossRef]
9. Verotti, M.; Crescenzi, R.; Balucani, M.; Belfiore, N.P. MEMS-based conjugate surfaces flexure hinge. *J. Mech. Des. Trans. ASME* **2015**, *137*. [CrossRef]
10. Belfiore, N.P.; Broggiato, G.; Verotti, M.; Balucani, M.; Crescenzi, R.; Bagolini, A.; Bellutti, P.; Boscardin, M. Simulation and construction of a mems CSFH based microgripper. *Int. J. Mech. Control* **2015**, *16*, 21–30.
11. Belfiore, N.P.; Broggiato, G.; Verotti, M.; Crescenzi, R.; Balucani, M.; Bagolini, A.; Bellutti, P.; Boscardin, M. Development of a MEMS technology CSFH based microgripper. In Proceedings of the 23rd International Conference on Robotics in Alpe-Adria-Danube Region, IEEE RAAD 2014, Smolenice Castle, Slovakia, 3–5 September 2014; Budinska, I., Havlik, S., Ciganek, J., Kozak, S., Hricko, J., Eds.; Institute of Electrical and Electronics Engineers Inc.: Piscataway, NJ, USA, 2015.
12. Crescenzi, R.; Balucani, M.; Belfiore, N.P. Operational characterization of CSFH MEMS technology based hinges. *J. Micromech. Microeng.* **2018**, *28*, 055012. [CrossRef]

13. Di Giamberardino, P.; Bagolini, A.; Bellutti, P.; Rudas, I.; Verotti, M.; Botta, F.; Belfiore, N. New MEMS tweezers for the viscoelastic characterization of soft materials at the microscale. *Micromachines* **2017**, *9*, 15. [CrossRef]

14. Bagolini, A.; Ronchin, S.; Bellutti, P.; Chistè, M.; Verotti, M.; Belfiore, N.P. Fabrication of Novel MEMS Microgrippers by Deep Reactive Ion Etching With Metal Hard Mask. *J. Microelectromech. Syst.* **2017**, *26*, 926–934. [CrossRef]

15. Dochshanov, A.; Verotti, M.; Belfiore, N. A Comprehensive Survey on Microgrippers Design: Operational Strategy. *J. Mech. Des. Trans. ASME* **2017**, *139*, 070801. [CrossRef]

16. Verotti, M.; Dochshanov, A.; Belfiore, N.P. A Comprehensive Survey on Microgrippers Design: Mechanical Structure. *J. Mech. Des. Trans. ASME* **2017**, *139*, 060801. [CrossRef]

17. Verotti, M.; Dochshanov, A.; Belfiore, N.P. Compliance Synthesis of CSFH MEMS-Based Microgrippers. *J. Mech. Des. Trans. ASME* **2017**, *139*, 022301 . [CrossRef]

18. Potrich, C.; Lunelli, L.; Bagolini, A.; Bellutti, P.; Pederzolli, C.; Verotti, M.; Belfiore, N.P. Innovative silicon microgrippers for biomedical applications: Design, mechanical simulation and evaluation of protein fouling. *Actuators* **2018**, *7*, 12. [CrossRef]

19. Sheets-Johnstone, M. Kinetic tactile-kinesthetic bodies: Ontogenetical foundations of apprenticeship learning. *Hum. Stud.* **2000**, *23*, 343–370. [CrossRef]

20. Mader, A.; Dertien, E. Tinkering as method in academic teaching. In Proceedings of the 18th International Conference on Engineering and Product Design Education, Aalborg, Denmark, 8–9 September 2016; pp. 240–245.

MDPI

St. Alban-Anlage 66

4052 Basel

Switzerland

Tel. +41 61 683 77 34

Fax +41 61 302 89 18

www.mdpi.com

Actuators Editorial Office

E-mail: actuators@mdpi.com

www.mdpi.com/journal/actuators

www.ingramcontent.com/pod-product-compliance
Lightning Source LLC
Chambersburg PA
CBHW051851210326
41597CB00033B/5859